Naturally Occurring Radioactive Material

Principles and Practices

Naturally Occurring Radioactive Material

Principles and Practices

PHILIP T. UNDERHILL

Advances in Environmental Science Series

T. Rick Irvin
Executive Editor

Camilo Cruz-Batres
Associate Editor

CRC Press
Taylor & Francis Group
Boca Raton London New York

CRC Press is an imprint of the
Taylor & Francis Group, an **informa** business

CRC Press
Taylor & Francis Group
6000 Broken Sound Parkway NW, Suite 300
Boca Raton, FL 33487-2742

First issued in paperback 2020

© 1996 by Taylor & Francis Group, LLC
CRC Press is an imprint of Taylor & Francis Group, an Informa business

No claim to original U.S. Government works

ISBN-13: 978-0-367-57960-9 (pbk)
ISBN-13: 978-1-57444-009-6 (hbk)

Visit the Taylor & Francis Web site at
http://www.taylorandfrancis.com

and the CRC Press Web site at
http://www.crcpress.com

TABLE OF CONTENTS

PREFACE

Radioactive materials have become a major environmental and workplace concern to a growing group of natural resource, medical, and manufacturing industries. Many mineral extraction and processing industries generate wastes that contain concentrated Naturally Occurring Radioactive Material (NORM). Recent attention has focused on the potential adverse environmental effects and human health hazards due to NORM. Several states have recently established regulations for the safe management of NORM. One of the fundamental tools for ensuring regulatory compliance and protecting personnel and the environment is training both management and staff in appropriate practices for NORM waste management, identification, and handling. *Naturally Occurring Radioactive Material (NORM): Principles and Practices* is intended to serve as a stand-alone review and course text for workers exposed to NORM and professionals responsible for NORM-related safety and waste management.

This text is also designed to support the course and credential programs in radiation safety and NORM management provided by the National Environmental Management and Education Center (NEMEC) and the National Registry of Environmental Professionals (NREP).

ACKNOWLEDGMENTS

I wish to thank Sandy Pearlman for her tireless and patient assistance in turning my text into a finished book (and me into an author!) and Camilo Cruz-Batres for his hard work and invaluable assistance with figures, reviewing, and editing.

Dr. Rick Irvin provided much-needed experience and occasional arm twisting to keep things moving along. Dennis Buda, President of St. Lucie

Press, has supported and guided the efforts of all concerned to bring this, the first volume in the *Advances in Environmental Science Series,* to press.

Special appreciation goes to Dick Young, Executive Director of NREP, who has guided the development of the credentialing and education programs for which this book was developed.

Finally, my wife, Jennifer, deserves much praise, simply for putting up with my authoring efforts.

AUTHORS

PHILIP T. UNDERHILL, B.S., RRPT, RRSCO, serves as Manager, Radiation Safety Programs for NEMEC (the National Environmental Management and Education Center). Mr. Underhill has developed numerous radiation safety courses for NEMEC, including courses for NORM workers, supervisors, and surveyors. He also chairs the National Registry of Environmental Professionals (NREP) examination committee for the Registered Radiation Safety Compliance Officer (RRSCO) credentialing program. He holds an honors degree in astrophysics from the University of London, England, and has over ten years of experience in the field of radiation protection as a member of management of oil and gas exploration/production firms as well as radiation safety and compliance consulting groups. He has developed and conducted radiation safety training and accreditation programs and has authored books, technical articles, and position papers on a wide range of environmental and workplace safety topics. Mr. Underhill is also the principal of P.T. Underhill and Associates, an environmental and radiation safety consulting practice. As a consultant, Mr. Underhill continues to provide radiation safety services to government agencies and petroleum, banking, refining, waste management, and legal firms.

DR. T. RICK IRVIN, Ph.D., REM, serves as Executive Director of NEMEC (the National Environmental Management and Education Center). NEMEC serves as the national academic, professional education, and credentialing center for the Environmental Industry Associations (EIA) and the NREP. He also currently chairs the NREP National Academic Advisory Board. Dr. Irvin received his B.S. degree *summa cum laude* from the University of Georgia and Ph.D. in toxicology from MIT in 1983. He served on the faculty of Texas A&M University from 1982 to 1990 and Louisiana State University from 1990 to 1995. The author of academic, industry, and government publications, he has served as Principal Investigator on over two million dollars of research funded by government, industry, and private foundations in the areas of environmental toxicology, chemical carcinogenesis and developmental toxicology, environmental site assessment and risk assessment, and risk-based monitoring and remediation technologies for waste management. Dr. Irvin has developed undergraduate, graduate, and industry short courses on environmental site assessment, environmental science and toxicology, and waste management. He serves as a consultant to environmental, financial, and chemical firms on the management of toxic wastes and hazardous materials. Dr. Irvin also serves as Executive Editor of the text series *Advances in Environmental Science* and *Advances in Environmental Management* published by St. Lucie Press.

CAMILO CRUZ-BATRES, M.S., REP, is a Senior Research Associate at NEMEC. He received his B.S. degree in microbiology in 1991 and master's degree in environmental studies in 1994 from Louisiana State University. He also served as a research assistant at the LSU Institute for Environmental Studies from 1993 to 1995. Mr. Cruz-Batres has conducted and directed validation of hazardous waste remediation technologies for complex mixtures of chemical wastes including wastes from the refining, oil production, and pulp and paper industries. He has also developed and validated laboratory toxicity test systems, employing *Ceriodaphnia* and *Photobacterium* sp. for identifying toxic constituents in hazardous waste mixtures. Mr. Cruz-Batres is currently serving as co-editor of *Microbial Toxicity Analysis of Environmental and Industrial Wastes.*

1   INTRODUCTION
TO **NORM**

STUDY OBJECTIVES

This chapter will enable the student to:

- Define the terms NORM and TENR.

- Discuss the origins of NORM in the oil industry and state where it is most often encountered.

- Name the three NORM radionuclides of primary concern in the oil and gas industry.

- Name the parent nuclides of radium and radon.

1.1 THE ORIGINS OF NORM

NORM is an acronym that stands for Naturally Occurring Radioactive Material. The radioactive elements from which NORM originates were incorporated into the earth's crust when the earth was created. There are many different types of radioactive material found in nature, but the oil and gas industry is only concerned with three particular radionuclides. They are:

Radium-226 (^{226}Ra)

Radium-228 (^{228}Ra)

Radon-222   (^{222}Rn)

Radium is of primary concern not only because it is radioactive, but also because it is chemically toxic. Radium may be almost as toxic as plutonium, the most toxic element known to man. (It is estimated that one teaspoon of plutonium could kill 100,000 people, due to its chemical toxicity alone.) Due to its chemical properties, radium is termed a "bone seeker." This means that when radium is ingested, it tends to collect in the bones of the body, where it stays for a very long time. It is for this reason that radium has been directly linked to leukemia and bone cancer.

Radon is somewhat different from radium. Radon is actually a radioactive gas. It is produced when radium undergoes radioactive decay. When it is in the gaseous state and is breathed in, 75 to 80% of the radon is exhaled with the next breath. However, the portion that remains in the lungs or in the bloodstream may undergo radioactive decay and change back into a non-gaseous form. This transformation allows it to remain in the lungs or elsewhere in the body for a long period of time. Radon is believed to be the leading cause of lung cancer after smoking.

Radium and radon are the parent nuclides of decay chains that contain up to 20 radioactive daughters. This means that radium and radon decay into products that are also radioactive, which in turn decay, etc. Radium and radon are therefore an even more significant hazard because their daughters will continue to irradiate body tissue even after the radium or radon has decayed.

1.1.1 WHERE DO RADIUM AND RADON COME FROM?

The origins of radium and radon are not immediately apparent (Figure 1.1). They are created as the result of the radioactive decay of other elements.

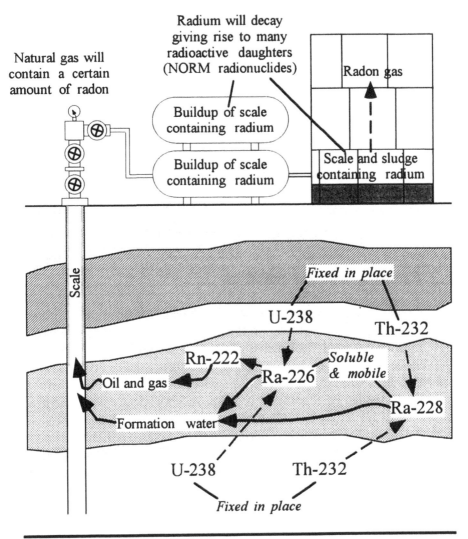

FIGURE 1.1 THE ORIGINS OF NORM

We refer to these elements as the parents of NORM or the hidden cause of NORM; they are:

Uranium-238 (^{238}U)

Thorium-232 (^{232}Th)

Uranium and thorium are the naturally occurring radioactive elements from which radium and radon are formed. They are known as primordial elements because they have always been present on earth. They were incorporated into the earth's crust at the time of its creation, more than 4.5 billion years ago, and they are slowly but continually decaying to radium and radon. There are surprisingly large amounts of these elements present on earth. If a 1-mile-by-1-mile square of soil, 1 foot thick, was processed and all the uranium and thorium were removed from it, an average of about 2.2 tons of uranium and 4.4 tons of thorium would be found. Because uranium and thorium are so plentiful, they produce large amounts of radium and radon.

Uranium-238 and thorium-232 each decay in a series of unique steps, known as their decay schemes, passing through a number of transformations until a stable or non-radioactive isotope is reached. The decay schemes for uranium-238 and thorium-232 are shown in Figure 1.2.

1.2 PRODUCTION OF NORM

From a regulatory standpoint, NORM is not being controlled unless it has been "technologically enhanced." Indeed, the acronym TENR (Technologically Enhanced Natural Radioactive material) is preferred by some people who (correctly) point out that any naturally occurring radioactive material, instead of just material which has been concentrated to the point where it becomes of regulatory concern, is NORM. Technological enhancement is considered to be the concentration of natural sources of radiation through some technical process which is not intended to produce radiation. This is the case with NORM in the oil and gas industry. The oil and gas production, refining, and storage processes which result in NORM are considered to be enhancements.

NORM becomes a problem when it is incorporated into the scale and sludge that are deposited inside equipment associated with oil and gas production. The deposition of NORM is usually associated with produced water. The reason for this is that radium comes out of the ground, dissolved in water.

The parents of radium-226 and radium-228, uranium and thorium, are found distributed throughout underground formations. The largest amounts are found in rock formations known to petroleum geologists as shales and other formations that contain some shale. Uranium and thorium are part of the matrix of the rock and as such are bound in place, being essentially insoluble in the reservoir fluids, which may be fresh water, salt water, oil, gas, or condensate. However, radium is somewhat soluble in water and is

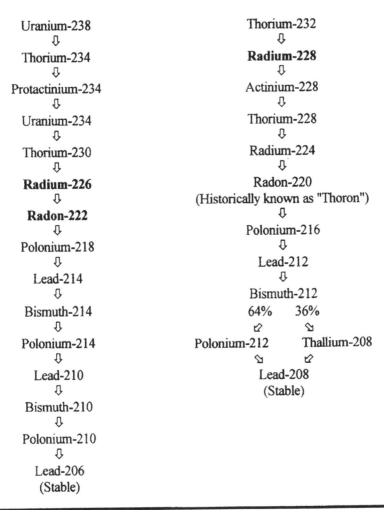

Uranium-238
⇩
Thorium-234
⇩
Protactinium-234
⇩
Uranium-234
⇩
Thorium-230
⇩
Radium-226
⇩
Radon-222
⇩
Polonium-218
⇩
Lead-214
⇩
Bismuth-214
⇩
Polonium-214
⇩
Lead-210
⇩
Bismuth-210
⇩
Polonium-210
⇩
Lead-206
(Stable)

Thorium-232
⇩
Radium-228
⇩
Actinium-228
⇩
Thorium-228
⇩
Radium-224
⇩
Radon-220
(Historically known as "Thoron")
⇩
Polonium-216
⇩
Lead-212
⇩
Bismuth-212
64%   36%
↗   ↘
Polonium-212   Thallium-208
↘   ↗
Lead-208
(Stable)

FIGURE 1.2   DECAY CHAINS OF URANIUM-238 AND THORIUM-232

therefore mobile. Consequently, radium may be produced along with any water. The concentration of radium which may be present in produced water will depend upon the amount and nature of shale in the formation and the physical and chemical conditions, such as pressure, temperature, and acidity (pH). Radium will be mixed (in very small proportions) with calcium, barium, and some strontium, which are much more abundant and possess similar chemical properties.

As produced fluids move from the reservoir to production facilities at the surface and then through various treatment processes, culminating at a petroleum refinery, the physical and chemical conditions change. This results in significant changes in the ability of the fluids to dissolve radium. If this ability decreases, then radium and the other elements with similar chemistry will come out of solution and will tend to be deposited as part of any scale or sludge.

Not all oil and gas wells produce NORM. NORM is rarely found without water production and, as a rule of thumb, NORM deposition is usually associated with *significant* volumes of water production. Unfortunately, because there are exceptions, this is not an absolute rule, but when consideration is given to the production history of a well which may have seen episodes of water production followed by the production of dry oil or gas, the rule of thumb usually holds true.

Not all NORM deposition is due to radium. Radon gas will be produced as a result of the decay of radium, either at the surface or in the formation. Radon can be produced either dissolved in produced fluids or mixed with natural gas. Gas plants can have very high radiation levels, especially where large volumes of gas are stored or compressed. Radon also poses a problem during the gas separation process at petroleum refineries as it all tends to come out of solution at once, when the lighter fluids such as ethane and propane are taken off the production stream.

There are a number of other NORM radionuclides, including uranium-235 (^{235}U) and its radioactive daughters and radioactive isotopes of potassium. However, these forms of NORM are either not normally encountered in the oil and gas industry or are generally in low concentrations relative to radium-226 and radium-228 and their daughters. An exception is potassium-40 (^{40}K), which is often found in elevated quantities. Potassium-40 poses a far smaller health risk than radium and consequently has a much higher exemption limit. It is rarely significant in NORM deposits, although it can be found in shale cuttings.

2 FUNDAMENTALS OF RADIOACTIVITY AND RADIOACTIVE MATERIALS

STUDY OBJECTIVES

This chapter will enable the student to:

- Describe the standard model of the atom and name its constituent particles.

- Describe the proton, neutron, and electron.

- Define atomic number and atomic mass number and how they relate to the physical, chemical, and radioactive properties of atoms.

- Understand the terms isotope and nuclide.

- Explain what the nuclear and coulomb forces are and the role their interaction plays in radioactive decay.

2.1 INTRODUCTION TO THE ATOM

Everything in the world around us is composed of atoms. There are now 112 different elemental types of atoms known to man. Of these, 92 are naturally occurring. The remainder can only be produced on earth by artificial means and are therefore referred to as "man-made." Each of these types of atoms represents a different element. An atom is the smallest indivisible part of an element that can take part in a chemical reaction.

Atoms combine to form molecules. A molecule is the smallest particle of a substance, made up of two or more atoms, that can express the definitive physical and chemical properties of that substance. A molecule need not contain atoms of more than one element.

Compounds are molecules formed of at least two different elements. The structure of a molecule or compound (i.e., the arrangement of its constituent atoms) defines the properties of that compound. Often, the properties of a compound will not resemble the properties of the elements that formed it.

2.2 ATOMIC COMPOSITION AND STRUCTURE

Atoms are normally considered to be made up of three different types of small (sub-atomic) particles. The standard, or Bohr, model describes the atom as a positively charged nucleus, around which circle small, light, negatively charged particles. The nucleus is composed of a tightly packed aggregation of two different particles called protons and neutrons, generically referred to as nucleons. Each proton carries a positive electrical charge, equal in magnitude to the negative charge of the orbiting particles which are known as electrons. (The magnitude of the electrical charge on an electron or proton is designated by the symbol e.) Neutrons have no electrical charge associated with them. Protons and neutrons are approximately the same size and mass.

The model of atomic structure illustrated in Figure 2.1 was postulated by the Danish physicist Niels Bohr to explain the known atomic properties. (Bohr received the Nobel Prize for Physics in 1922, and his son, also an atomic physicist, was similarly honored in 1975.)

The size of the electron's orbits is very large when compared to the size of the nucleus. On a relative scale, if the nucleus were the size of a baseball,

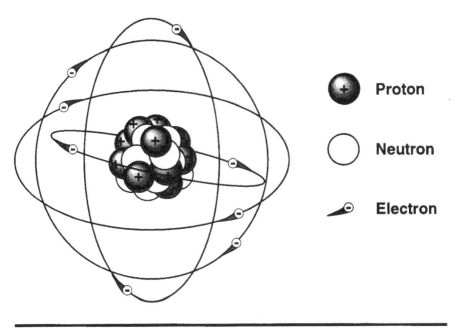

FIGURE 2.1   BOHR MODEL OF THE ATOM

the electrons would be the size of grains of sand and the nearest electron would be several hundred yards from the nucleus.

Protons and neutrons are very massive (heavy) compared to electrons. Because of this, the nucleus contains approximately 99.97% of the mass of the atom. All atoms will tend toward the stable state of electrical neutrality, where the atom has no net electrical charge. Thus, most atoms will have equal numbers of protons and electrons.

The properties of the three sub-atomic particles are listed in Table 2.1.

2.3 ATOMIC PROPERTIES

An atom may be described uniquely, in terms of the numbers of protons, neutrons, and electrons of which it is comprised. All the classical properties of atoms with the same corresponding numbers of protons, neutrons, and electrons will be exactly identical.

It is convenient to describe three groups of classical properties for any atom:

TABLE 2.1   PROPERTIES OF SUB-ATOMIC PARTICLES

PROTON	• Located in the nucleus • Single, positive electrical charge of magnitude +e • Rest mass of 1.673×10^{-24} grams • Designated by the symbol p or p^+
NEUTRON	• Located in the nucleus • Electrically neutral (no charge) • Rest mass similar to a proton at 1.675×10^{-24} grams • Symbol is n
ELECTRON	• Orbits around the nucleus • Single, negative electrical charge of magnitude –e • Rest mass is $1/_{1836}$ that of a proton or 9.101×10^{-28} grams • Notation is e or e^-

PHYSICAL PROPERTIES	Such as what element does an atom represent and what is the average density, melting point, electrical conductivity, etc. of a macroscopic sample of the element?
CHEMICAL PROPERTIES	What type of reactions does the atom typically enter into and how vigorously does it do so?
RADIOACTIVE PROPERTIES	What type(s) of radiation is emitted when an atom undergoes radioactive decay, how much energy is released, and how likely is it for a given atom to decay within a specific time period?

In order to understand how the presence of differing numbers of each of the sub-atomic particles can affect the overall properties of an atom, certain terms need to be defined and discussed.

ATOMIC NUMBER	The number of protons in the nucleus of an atom. Represented by the symbol **Z**. Sometimes known as the proton number of an atom.
ATOMIC MASS NUMBER	The total number of nucleons, i.e., the sum of the number of protons and neutrons in the nucleus. Represented by the symbol **A**. Sometimes known as the nucleon number or mass number of an atom.

2.3.1 PHYSICAL PROPERTIES OF THE ATOM

By convention, the number of protons in the nucleus of an atom determines what element it is. This makes sense, because the number of protons determines the number of electrons and thus the chemical properties of the atom. Historically, all elements were known by their chemical properties. Any atoms with the same specific number of protons in their nuclei will always be the same element, regardless of the number of neutrons present.

Utilizing the preceding definitions, it could be said that the atomic number of an atom defines which element it is. Atoms with the same atomic number will always be the same element, regardless of their atomic mass number.

The atomic mass number of an atom should not be confused with atomic mass, which is the actual mass of the atom, commonly measured in atomic mass units. One atomic mass unit (amu) is equal to one-twelfth of the mass of an atom of carbon-12 (^{12}C) or 1.66×10^{-24} grams. An alternative term for the amu is the dalton, named after English chemist John Dalton, who formulated the modern version of the atomic theory.

2.3.2 CHEMICAL PROPERTIES OF THE ATOM

According to the Bohr model, electrons circle the nucleus in "orbits" known as valence shells. The complex theories of quantum mechanics describe the shape and relative orientation of these shells. One of the consequences of the quantum nature of electrons orbiting an atom is that only a certain number of electrons can occupy each shell. Once a shell is filled with its quota of electrons, the next electron will be alone in the next shell. Other atoms "looking in" at an atom can only "see" the outermost shell. Thus, the number of electrons in the outermost shell of an atom is fundamental in defining the way in which it can react with other atoms. The total number of electrons is less important, although it does affect the intensity with which the chemical properties are expressed. It can therefore be stated that atoms with an equal number of electrons in their outermost shells will possess similar chemical properties. This is the principle upon which the layout of the Periodic Table of the Elements (Figure 2.2) is based.

2.3.3 RADIOACTIVE PROPERTIES OF THE ATOM

Isotopes are atoms that have the same number of protons but different numbers of neutrons. Because they have different numbers of neutrons, they will also have different atomic mass numbers. Therefore, it can be said that

Elements in the same row (Arabic Numerals) occupy the same PERIOD
Elements in columns (Roman Numerals & letters) occupy the same GROUP

Legend:
- Noble Gasses
- Halogens
- Non Metals
- Metals
- Alkali Metals
- Transition Metals
- Rare Earth Elements
- Normal - Naturally occurring
- Underline - Artificially prepared
- X ← Element Symbol
- Z ← Atomic Number

	IA	IIA	IIIA	IVA	VA	VIA	VIIA	VIIIA	VIIIA	VIIIA	IB	IIB	IIIB	IVB	VB	VIB	VIIB	VIII
1	H 1																	He 2
2	Li 3	Be 4											B 5	C 6	N 7	O 8	F 9	Ne 10
3	Na 11	Mg 12											Al 13	Si 14	P 15	S 16	Cl 17	Ar 18
4	K 19	Ca 20	Sc 21	Ti 22	V 23	Cr 24	Mn 25	Fe 26	Co 27	Ni 28	Cu 29	Zn 30	Ga 31	Ge 32	As 33	Se 34	Br 35	Kr 36
5	Rb 37	Sr 38	Y 39	Zr 40	Nb 41	Mb 42	_Tc_ 43	Ru 44	Rh 45	Pd 46	Ag 47	Cd 48	In 49	Sn 50	Sb 51	Te 52	I 53	Xe 54
6	Cs 55	Ba 56	La 57	Hf 72	Ta 73	W 74	Re 75	Os 76	Ir 77	Pt 78	Au 79	Hg 80	Tl 81	Pb 82	Ti 83	Po 84	At 85	Rn 86
7	Fr 87	Ra 88	Ac 89	_Rf_ 104	_Ha_ 105	? 106	? 107	? 108	? 109	? 110	? 111	? 112						

Lanthanide Series	Ce 58	Pr 59	Nd 60	_Pm_ 61	Sm 62	Eu 63	Gb 64	Td 65	Dy 66	Ho 67	Er 68	Tm 69	Yb 70	Lu 71
Actinide Series	Th 90	Pa 91	_U_ 92	_Np_ 93	_Pu_ 94	_Am_ 95	_Cm_ 96	_Bk_ 97	_Cf_ 98	_Es_ 99	_Fm_ 100	_Md_ 101	_No_ 102	_Lr_ 103

FIGURE 2.2   PERIODIC TABLE OF THE ELEMENTS

isotopes are atoms with the same atomic number but different atomic mass numbers. Different isotopes of an element will have near identical chemical properties but differing physical properties (including different radioactive properties).

For example, the element sulfur is characterized by having 16 protons in its nucleus, giving it (by definition) an atomic number of 16. However, sulfur atoms have been known to have any number of neutrons, from 13 to 28. This means that an atom of sulfur can have an atomic mass of between 29 and 44. All these different sulfur atoms are known as isotopes of sulfur.

Four of the isotopes of sulfur are stable or non-radioactive (^{32}S, ^{33}S, ^{34}S, and ^{36}S). These isotopes are all found in varying abundances in naturally occurring sulfur, while the rest are unstable or radioactive. Most of these can be found only in an atomic laboratory, as they will exist for only a few hours, minutes, or seconds.

The element hydrogen also possesses several isotopes. Hydrogen is the simplest atom. In its most abundant atomic form, its nucleus consists of a single proton. An isotope of hydrogen, called deuterium, has one proton but also contains a neutron. Normal hydrogen has an atomic mass number of one, whereas deuterium has an atomic mass number of two. Tritium, the third isotope of hydrogen, has one proton and two neutrons. Its atomic mass number is three. Hydrogen and deuterium are stable, whereas tritium is radioactive. The three isotopes of hydrogen are illustrated in Figure 2.3.

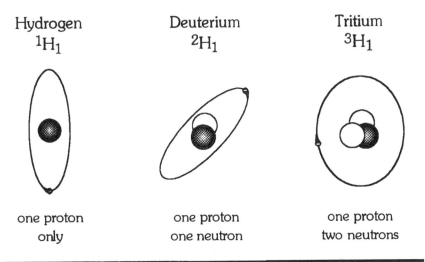

| Hydrogen | Deuterium | Tritium |
| $^{1}H_{1}$ | $^{2}H_{1}$ | $^{3}H_{1}$ |

one proton only   one proton one neutron   one proton two neutrons

FIGURE 2.3 ISOTOPES OF HYDROGEN

By definition, isotopes are atoms of the same element. Atoms of different elements characterized by their atomic number, atomic mass number, and nuclear energy state are referred to as nuclides. Radioactive nuclides are frequently referred to as radionuclides.

Standard conventional nomenclature for describing nuclides utilizes superscript and subscript notations:

$$^{A}X_{Z}$$

where Z and A are the atomic number and atomic mass number, respectively, and X is the symbol for the element. The following are examples of several nuclides annotated in this manner:

$$^{56}Fe_{26} \text{ (Iron-56)}$$

$$^{3}H_{1} \text{ (Tritium)}$$

$$^{14}C_{6} \text{ (Carbon-14)}$$

Because this convention is not quite universal, sometimes both superscript and subscript may precede the symbol for the element and sometimes the superscript may follow it.

2.4 THE NUCLEAR FORCE

The nucleus of an atom consists of neutrons and protons tightly packed together. Because protons are all positively charged, they are mutually repelled from each other. This process, which is very much like trying to put like poles of two magnets together in that they push apart, is called electrostatic or coulombic repulsion. Some force must act to overcome this repulsive coulomb force, or otherwise the nucleus would fly apart. This force is called the nuclear force.

The nuclear force acts between the particles of the nucleus, binding it together. The nuclear force has quite different properties than the coulomb force, much in the same way as the familiar forces of gravity and electromagnetism are different. Some of the distinctive properties of the nuclear force are:

- It is charge independent. This means that it is just as strong regardless of whether it is acting between charged or uncharged nucleons, i.e., protons or neutrons.

- It is very strong (approximately 137 times stronger than the coulomb force).

- It has a very short effective range ($<10^{-15}$ meters). This is somewhat shorter than the diameter of a typical nucleus (1.5×10^{-15} to 9×10^{-15} meters) and many orders of magnitude smaller than the effective range of the coulomb force.

The explanation of exactly how the nuclear force leads to the phenomenon of radioactivity is very complicated and involves fiendish mathematics and advanced concepts in physics. However, the basic principle has to do with the combined effect of the "struggle" between the repulsive coulombic force and the attractive nuclear force. The two opposing forces are unequal in magnitude, direction, and distance of action, yet, most of the time, the much stronger nuclear force "wins" and the nucleus of an atom stays intact. However, the nucleons in an atomic nucleus are not fixed in place; they are constantly jostling and rattling around. Occasionally, one or more of these nucleons may be jostled just far enough to be affected more by the coulombic force than the nuclear force. When this happens, the particle (or particles) will be expelled from the nucleus as radiation. The extent of the imbalance between the two forces is related to the number of protons and the total number of nucleons. For any given element, therefore, the radioactive properties will depend upon how many neutrons are present.

Radium demonstrates this admirably. Radium-226, with 88 protons and 138 neutrons, is an alpha emitter with a half-life of 1600 years, whereas radium-228, which has 88 protons and 140 neutrons, is a beta emitter and has a half-life of only 5.76 years.

3 ORIGIN AND CLASSIFICATION OF RADIATION

STUDY OBJECTIVES

This chapter will enable the student to:

- Define radiation.

- List the three main types of radiation emitted by NORM and relate their properties.

- Understand the atomic decay mechanisms that produce each form of radiation.

- Identify and use appropriate units for measuring radiation exposure, biological dose equivalent, and radioactive material.

- Understand the concept of half-life as it relates to the radioactive decay and biological elimination of radioactive materials.

3.1 RADIATION

Generally, radiation can be defined as particles or energy released by an unstable atom. An unstable atom has an arrangement of nucleons with sufficient excess energy to cause decay by emission of radiation. The resulting nuclide will be left with less energy.

It is important to understand the distinction between particulate and non-particulate radiation. Radiation in the form of a particle will have a measurable size, mass, and velocity, whereas non-particulate radiation is pure energy traveling at the speed of light.

Dozens of types of radiation are known to science. Radiation is classified according to the nature of the particles emitted (if any), the amount of energy associated with their emission, and the mechanism by which the emission arises. Most of these types of radiation are of interest only to high-energy and nuclear physicists. The radiation emissions most commonly encountered by persons working in the oil industry are gamma rays, alpha particles, and beta particles. Oil well loggers may use special radioactive sources which give off neutron radiation, but this is not of any special interest where NORM is concerned.

The unit used to quantify the amount of energy released by a nuclear reaction is called the electron volt (eV). The energy gained by an electron when accelerated through a potential difference of one volt is equal to one electron volt. Because the electron volt is such a small unit, it is not convenient to use when describing the energy released by nuclear decay. The terms kilo electron volt (keV) and mega electron volt (MeV) are commonly used in this case.

$$1 \text{ keV} = 1,000 \text{ eV} = 1 \times 10^3 \text{ eV}$$

$$1 \text{ MeV} = 1,000,000 \text{ eV} = 1 \times 10^6 \text{ eV}$$

3.2 GAMMA RADIATION (SYMBOL: γ)

Gamma radiation is excess energy that is released from the nucleus of an atom. Gamma rays have no mass or electrical charge. They are a form of electromagnetic radiation, similar to visible light, and hence the common terminology gamma ray. However, they are often considered to be small packets of energy known as photons. Indeed, gamma rays display some of the properties of both waves and massless particles traveling at the speed

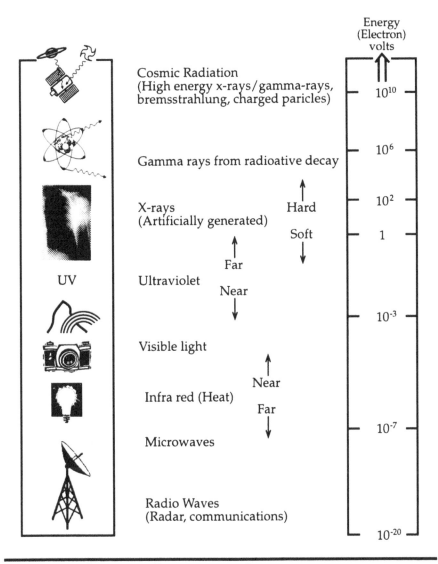

FIGURE 3.1   SPECTRUM OF ELECTROMAGNETIC RADIATION

of light, which leads to the concept of wave/particle duality. Gamma rays and other forms of electromagnetic radiation are shown in Figure 3.1.

X-rays are also electromagnetic energy waves. For all practical purposes, x- and gamma rays are identical. The distinction is made to differentiate between their origins. X-rays originate when excited electrons give up some excess energy. They tend to have lower energies than gamma rays. Photon radiation resulting from charged particles undergoing collisions or acceleration is also called x-ray radiation or bremsstrahlung radiation. It may have energies far in excess of normally encountered gamma ray energies.

Gamma rays are considered to have a "whole body" penetrating ability because they can travel a considerable distance in body tissue. They can also pass through the considerable thickness of most commonly encountered materials including wood, concrete, and to a certain extent steel and lead, which is why gamma ray exposure rate surveys are useful in general screening for the presence of NORM.

Gamma rays are emitted with distinctive energy levels or quanta which are usually unique to the atomic transition that causes their emission (i.e., the radioactive decay of a particular radionuclide will always produce gamma rays with the same, uniquely recognizable energy). This allows identification of gamma-emitting radionuclides by a technique known as gamma spectroscopy.

As previously stated, the emission of radiation is associated with the decay of an unstable atomic nucleus into a nucleus with a lower energy state. However, the new nucleus may still have sufficient instability or excess energy to decay again or emit additional radiation. Gamma radiation is usually emitted in this fashion, as illustrated in Figure 3.2.

Although the gamma ray emissions commonly associated with many beta and all alpha emissions are usually considered to happen simultaneously to the particle emission, there is often a very short delay. The atom is in an excited (excess energy) state between these two emissions, which is why the gamma ray is given off in the first place. If the intermediate excited state is not so unstable that the gamma ray is virtually instanta-

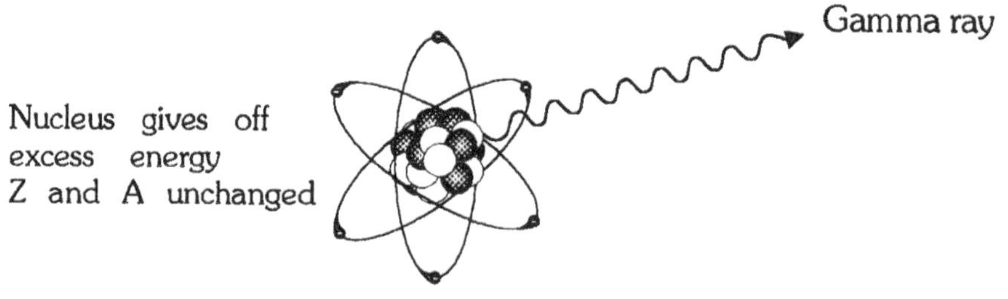

Gamma ray

Nucleus gives off excess energy Z and A unchanged

FIGURE 3.2 GAMMA RADIATION

neous, the nucleus is considered to be in a metastable state. In such circumstances, the gamma emission, when it happens, may be considered to be a separate radioactive event in its own right. Metastable atoms rarely occur in nature. They are designated by adding a lower-case letter "m" to their isotopic designation. The man-made atom technetium-99 (^{99m}Tc) is the most commonly encountered metastable radionuclide.

3.3 ALPHA PARTICLES (SYMBOL: α)

An alpha particle is literally a part of an atomic nucleus that is expelled during radioactive decay. An alpha particle consists of two neutrons and two protons together. This gives it a characteristic atomic mass number of four. It also has a net electrical charge of +2e associated with it. Examining the periodic table will show that an alpha particle is basically a helium atom without its orbital electrons. Atomic decay with emission of an alpha particle is illustrated in Figure 3.3.

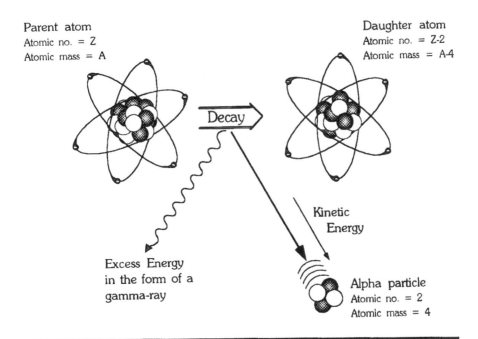

Parent atom
Atomic no. = Z
Atomic mass = A

Daughter atom
Atomic no. = Z-2
Atomic mass = A-4

Decay

Kinetic
Energy

Excess Energy
in the form of a
gamma-ray

Alpha particle
Atomic no. = 2
Atomic mass = 4

FIGURE 3.3   ALPHA DECAY

Because of their high mass and electrical charge, alpha particles interact with matter very readily, and their length of travel in air and organic tissue is very short. Alpha particles can only travel a few centimeters in dry air (less in humid air) and can be shielded by a sheet of paper. Due to their high mass, charge, and energy, alpha radiation has a very high ability to ionize tissue and cause biological damage.

Alpha particles have discrete energy levels associated with them. This means that alpha spectroscopy may be performed to identify a particular alpha-emitting isotope by the characteristic energy levels of its emissions.

3.4 BETA PARTICLES (SYMBOL: β)

Beta radiation is a particulate form of radiation. Beta particles are high-energy electrons that have originated in the nucleus of an atom, rather than in the electron shells that surround the nucleus. They can be considered to be the result of an unstable neutron turning into a proton plus an electron, although it is actually somewhat more complicated than that. The mass of a beta particle is (not surprisingly) the same as that of an electron or approximately $1/1836$ that of a neutron or proton.

Typically, a beta particle has a single, negative electrical charge associated with it. In some instances, however, beta particles can have a positive charge. Such particles are the anti-matter equivalents of electrons or beta particles. They are known as positrons. Positrons rarely occur in association with natural radioactive decay. There are no positron emissions associated with NORM.

Beta emissions have no specific energies by which they may be uniquely identified in the manner of gamma or alpha radiation. Therefore, there is no such thing as beta spectroscopy. The reason for this is that a beta particle is always emitted in conjunction with a particle known as an anti-neutrino (symbol: $\bar{\upsilon}$). The two particles emitted share the total energy of the decay in varying, unpredictable proportions. The anti-neutrino reacts so weakly with matter that it is almost impossible to detect. Beta decay is depicted in Figure 3.4.

3.5 PROPERTIES OF RADIATION

The significant properties of gamma, alpha, and beta radiation are summarized in Table 3.1.

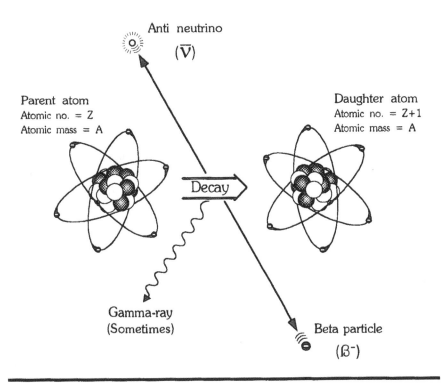

FIGURE 3.4 BETA DECAY

TABLE 3.1 PROPERTIES OF GAMMA, ALPHA, AND BETA RADIATION

Radiation type ⟹	Gamma (rays)	Alpha (particles)	Beta (particles)
Atomic mass number	Zero	4	$1/1830$
Electrical charge	Zero	+2e	−e
Typical energy	0.6 MeV	6 MeV	1 MeV
Discrete energy levels	Yes	Yes	No
Penetrating ability	6 inches of lead "Whole body"	Stopped by dead layer of skin	Stopped by thick cardboard
Travel in dry air	N/A*	1 inch	10 feet
Relative ionizing ability	1	40,000	100

* N/A = not applicable.

3.6 MODES OF DECAY

The act of a nucleus undergoing radioactive decay to emit a particle (alpha or beta radiation) and thereby changing into a new nucleus is often known as nuclear disintegration or, occasionally, transmutation. Some unstable nuclides can decay in more than one fashion (e.g., by either alpha or beta emission or by alpha and beta occurring essentially simultaneously). These are known as the different modes of decay for that nuclide.

Often radionuclides may decay by only one mode, but the energy of the emitted particle or photon can take any one of several values. In cases where the nuclide has more than one decay mode, each mode can have many energies associated with it. Because of these facts, a person trying to count disintegrations from a particular radionuclide may have to look for more than one type of radiation at many different energies. The relative proportions of atoms that follow any particular "route" of decay are fixed relative to each other. For example, bismuth-212, a radionuclide in the natural thorium-232 decay chain, can decay by alpha or beta emission. A percentage (36%) of atoms decay by the alpha mode and the remainder by the beta mode. Each of the two daughter radionuclides decays by the opposite mode to that which formed it; thus, the ultimate daughter product is the same. The decay schematic for bismuth-212 is illustrated in Figure 3.5.

Each of the four different atomic decays shown in Figure 3.5 has several different possible energies associated with it. Additionally, as indicated, each decay (whether beta or alpha) also has an associated gamma ray emission. Each of these gamma rays can have one of several different, unique energies.

3.7 MEASURING RADIATION

Most of the scientific world measures radiation and radioactivity using the Système Internationale (S.I.) units of measurement. This system for radiation measurement is beginning to be adopted in the United States, but many of the old units are still commonly used. Following is an explanation of units used to measure several different radiation-related quantities. Traditional units are listed first, followed by their S.I. counterparts.

ROENTGEN (OR RÖNTGEN), SYMBOL: **R**—The amount of gamma or x-ray radiation that will produce, in dry air, a charge of 2.58×10^{-4} coulombs on

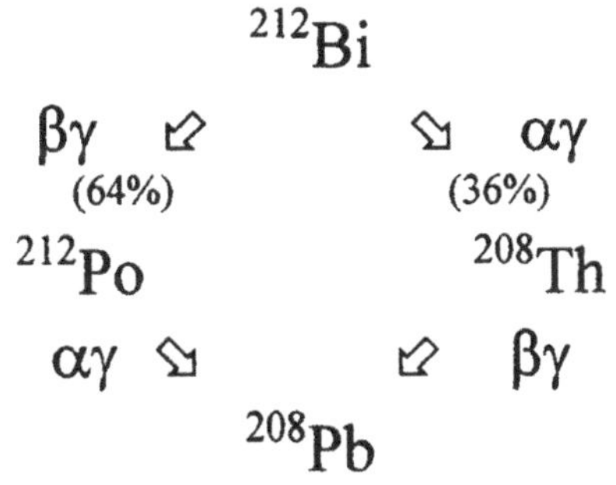

$$^{212}\text{Bi}$$

βγ ↙   ↘ αγ
(64%)   (36%)

$$^{212}\text{Po} \qquad\qquad ^{208}\text{Th}$$

αγ ↘   ↙ βγ

$$^{208}\text{Pb}$$

Bi - Bismuth
Po - Polonium
Th - Thorium
Pb - Lead

FIGURE 3.5 DECAY SCHEMATIC OF BISMUTH-212

all the ions of one sign, when all the electrons released in a volume of air, of mass 1 kilogram, are completely stopped.

$$1 \text{ R} = 2.58 \times 10^{-4} \text{ C kg}^{-1}$$

This means that the roentgen is a measurement of the ability of radiation to ionize air. Therefore, the roentgen is a unit of exposure and not dose. The unit can only be correctly applied to gamma or x-rays with energies that are greater than 32 electron volts and less than 4 mega electron volts. The reason for the lower limit is that a minimum of 32 electron volts is required to ionize air. The upper limit is less well defined but is basically the crossover point at which the mechanism by which photons give up energy changes and the definition of the roentgen becomes invalid.

The unit is named after Wilhelm Conrad Roentgen, a professor of physics at the University of Würzburg, who discovered x-rays in 1895. The discovery was made accidentally when experimenting with an early cathode ray tube. In 1901, Roentgen became the first recipient of the Nobel Prize for Physics.

There is no equivalent to the roentgen in the S.I. system. Instead, radiation measurements are usually reported in terms of absorbed dose. The roentgen was redefined in more easily measured terms to fit with the unified approach to S.I. units. The new definition is 2.58×10^{-4} coulombs per kilogram of air.

Exposure is a measure of the field of radiation to which a person or object is subjected. An object may be exposed to a large field of radiation, but only a small portion of it may interact with the object. Dose is a measurement of how much of the radiation has interacted with matter.

The roentgen gives a measurement of the amount of radiation an object or body has been exposed to, but not how much energy has been deposited or how much biological damage has been done. The rad (Radiation Absorbed Dose) was established as a unit to measure ionization in any material.

RAD (RADIATION ABSORBED DOSE)—The amount of any type of radiation that will deposit 0.01 joules per kilogram (100 ergs per gram) of energy in any material. This makes the rad a unit of energy deposition.

The S.I. equivalent to the rad is the gray, named after English physicist Louis Gray.

GRAY (GY)—Unit of absorbed dose: 1 Gy = 1 joule per kilogram

$$1 \text{ Gy} = 100 \text{ rad}$$

$$(1 \text{ rad} = 0.01 \text{ Gy})$$

(Strictly, the rad and gray can be considered units of dose, but this is not encouraged due to the potential for confusion with measurement of biological dose equivalent.)

Despite being the preferred unit of human dose for many years (and still in common usage on "Star Trek"), the rad is no longer used for measuring radiation dose to an organism because it does not indicate how much biological damage has actually occurred. A unit called the rem (Roentgen Equivalent Man) was developed to equate deposited energy to biological damage (biological dose equivalent). The rem and its S.I. equivalent, the sievert, are the units of choice for radiation protection.

REM (ROENTGEN EQUIVALENT MAN)—The amount of any type of radiation that will cause damage to body tissue equivalent to the damage that would be caused by depositing 100 ergs of gamma radiation per gram of body tissue. The rem is a measurement of biological dose equivalent.

Rem may be approximated by multiplying rad by a quality factor:

$$\text{rem} = \text{rad} \times \text{Q.F.}$$

TABLE 3.2   TYPICAL QUALITY FACTORS FOR NATURALLY OCCURRING RADIATION

Radiation type $\Rightarrow$	Gamma	Beta	Alpha
Quality factor	1	1	≈ 20

The quality factor is dependent upon the type of radiation, its energy, and the tissues with which it is reacting. Typical quality factors for naturally occurring radiation are presented in Table 3.2.

When radiation exposure limits were first established, they were actually dose limits. One limit that was used in the medical community was the dose that would cause a pronounced reddening of the skin called erythema. This was called the erythema dose.

The S.I. equivalent to the rem is the sievert, named in honor of Swedish mathematician and physicist Rolf Sievert. Sievert was a pioneer in the development of radiation protection standards.

SIEVERT (Sv)—Unit of human dose equivalent: 1 Sv = 100 rem = 10^5 mrem

$$(10^{-5} \text{ Sv} = 1 \text{ mrem})$$

$$(0.01 \text{ Sv} = 1 \text{ rem})$$

3.8 MEASURING RADIOACTIVITY

Radioactivity is fundamentally different from radiation. The radioactivity of a substance is a measure of the rate at which radioactive decays are taking place. Radioactivity is quantified using a unit called a curie. Therefore, the curie is a basic unit of activity measurement.

CURIE—The amount of any radioactive material that will decay at the rate of 3.7×10^{10} disintegrations per second (dps) or 2.22×10^{12} disintegrations per minute (dpm).

The curie was named after Pierre and Marie Curie. Working in their Paris laboratory around the turn of the century, the Curies isolated and measured the properties of several new radioactive elements, including radium. The Curies received the Nobel Prize for Physics in 1903. In 1911, Marie Curie again received a Nobel Prize for Chemistry for her work in isolating the new elements. She died in 1934 of diseases related to excessive radiation exposure. Activity measurements were initially based on radium because of its popularity among scientists experimenting with radiation and because of its long half-life. In fact, the curie is based upon the radioactivity of one gram of radium-226 in equilibrium with its daughter radionuclides.

When the Curies received their first Nobel Prize, they shared it with Antoine Becquerel, a fellow Frenchman. Becquerel was the discoverer of natural radioactivity and postulated its atomic origin. As did Roentgen, Becquerel made his discovery accidentally when he left some uranium salts near a photographic plate which was exposed by the radiation. Becquerel and the Curies also share the nomenclature of units for measuring radioactivity. The S.I. equivalent of the curie is the becquerel.

BECQUEREL (BQ)—Unit of activity: 1 Bq = 1 disintegration per second

$$1 \text{ Bq} = 2.7 \times 10^{-11} \text{ Ci} = 27 \text{ pCi}$$

$$(3.7 \times 10^{10} \text{ Bq} = 1 \text{ Ci})$$

$$(0.037 \text{ Bq} = 1 \text{ pCi})$$

3.9 MODIFYING UNITS: STANDARD SCIENTIFIC NOTATION

Depending upon the application, required measurements of radiation and radioactive material can cover an enormous range of many orders of magnitude. For example, environmental monitoring for radionuclides in water may result in measurements of a few trillionths of a curie or hundredths of a becquerel. Conversely, industrial radiographers may use sealed sources with an activity in excess of 100 curies or approximately 4 terabecquerels. To overcome the need for exponential scientific notation or endless zeros and decimal places, Greek suffix notation is used. The commonly used notation symbols and their respective multiplicative factors are displayed in Table 3.3.

Radiation and radioactive measurements are rarely reported without appropriate modifiers. Some of the most common units and their uses are outlined in Table 3.4.

Measurements of absolute activity are used for sealed sources, such as those used for well logging, the x-ray of welds, and other non-destructive testing. When making measurements of NORM radioactivity, specific activities or concentrations are usually measured. Specific activities are activities per unit mass or per unit volume.

TABLE 3.3 Standard Scientific Notation for Modifying Units

Prefix		Multiplicative Factor	
Symbol	Name	Decimal Notation	Scientific (Exponential) Notation
T	tera	1,000,000,000,000	10^{12}
G	giga	1,000,000,000	10^{9}
M	mega	1,000,000	10^{6}
K	kilo	1,000	10^{3}
m	milli	0.001	10^{-3}
μ	micro	0.000,001	10^{-6}
n	nano	0.000,000,001	10^{-9}
p	pico	0.000,000,000,001	10^{-12}
f	femto	0.000,000,000,000,001	10^{-15}

TABLE 3.4 Commonly Used Units and Their Equivalents

Measurement	Common Units		
	Longhand	Symbol	Equivalent
Radiation exposure rate	microroentgen per hour	μR/hr	none
Absorbed dose	rad	rad	10 mGy
	milligray	mGy	0.1 rad
	millirad	mrad	10 μGy
	microgray	μGy	0.1 mrad
Equivalent dose	millirem	mrem	10 μSv
	microsievert	μSv	0.1 mrem
Activity (radioactivity)	millicurie	mCi	37 MBq
	megabecquerel	MBq	27 μCi
	microcurie	μCi	37 kBq
	kilobecquerel	kBq	27 nCi
	nanocurie	nCi	37 Bq
	becquerel	Bq	27 pCi
	picocurie	pCi	37 mBq

Common uses for specific activity units are:

Measurement of radioactive material in air:	microcuries per milliliter (μCi/ml)
	or becquerels per cubic meter (Bq/m^3)
Quantifying radioactive material on surfaces:	microcuries per square centimeter (μCi/cm^2)
	or becquerels per square meter (Bq/m^2)
Measuring radioactive material in liquids:	microcuries per milliliter (μCi/ml)
	or becquerels per cubic meter (Bq/m^3)
	or picocuries per liter (pCi/l)
	or picocuries per milliliter (pCi/ml)
Measuring concentration in soil or solids:	picocuries per gram (pCi/g)
	or becquerels per kilogram (Bq/kg)

3.10 HALF-LIVES OF RADIOACTIVE MATERIALS

3.10.1 RADIOACTIVE HALF-LIFE

Every type of radioactive material undergoes radioactive decay at a certain rate that is virtually unaffected by changes in temperature, pressure, magnetic field, or other external influences. The rate of decay is proportional to the number of the original nuclides present, which results in an exponential decrease, over time, in the amount of that material. Ignoring the contribution due to its radioactive daughters, the radioactivity of a sample of radioactive material containing a single radionuclide will always decrease at the same rate.

Thus, the time taken for one-half of the original nuclides to undergo radioactive decay is equal to the time taken for their radioactivity to decay to one-half the original value. This length of time is called the half-life ($T_{1/2r}$) of the radionuclide.

The fraction of atoms decaying in a given time is not truly a constant because radioactivity is a statistical process. Also, any one type of unstable nuclide may (and indeed typically will) have more than one possible mode of decay with different probabilities and therefore different half-lives. However, because an external observer is unable to significantly influence which decay should take place, the statistics of the process lead the observer to

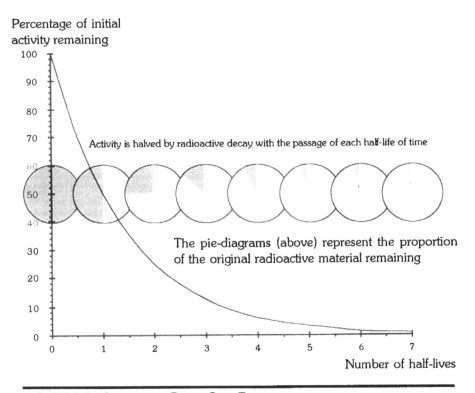

Percentage of initial
activity remaining

Activity is halved by radioactive decay with the passage of each half-life of time

The pie-diagrams (above) represent the proportion
of the original radioactive material remaining

Number of half-lives

FIGURE 3.6   RADIOACTIVE DECAY OVER TIME

measure a fixed half-life with fixed relative proportions of each of the possible decay modes. Due to the large number of atoms usually involved, the half-life appears to be a constant and can be measured as such with considerable accuracy.

In order to avoid confusion with other terms used in the biological and chemical sciences and radiation protection, the half-life of a radioactive isotope, as described above, is termed the radioactive half-life ($T_{1/2r}$).

The radioactive half-lives for radium-226 and radium-228 are 1608 and 5.76 years, respectively. The radioactive decay of radium-226 over a period of 8000 years is shown in Figure 3.6. It should be obvious that the overall activity decreases fairly rapidly once a few half-lives have passed. Depending upon the initial activity, common practice is to assume that a sample of any radioactive material has decayed away to essentially background levels in seven to ten half-lives. For radium-226, this would be between 11,000 and 16,000 years.

3.10.2 BIOLOGICAL HALF-LIFE

Radionuclides deposited within the human body are usually considered to be eliminated from the body at a rate proportional to the amount of the radionuclide present, resulting in an exponential decrease in the amount of the material in the body. This process of elimination is analogous to the radioactive half-life and is termed the biological half-life ($T_{1/2b}$). The biological half-life is the approximate time taken for an average adult human to eliminate one-half of a radioactive substance, by normal biological means only. This parameter is calculated independent of the radioactive properties of the substance and is determined strictly by the ability of the body to remove the particular element (e.g., radium, iodine, etc.).

The concept of partitioning assigns elimination half-lives to a wide range of tissues in the body. By making many assumptions concerning transportation and elimination systems that operate between different tissues, a complicated series of calculations allows the determination of half-lives for particular organs. It should be noted that such half-lives are highly dependent upon the element in question. For example, any of the several isotopes of iodine are readily retained in the thyroid yet rapidly eliminated from most other tissues. In organisms other than man, differences in metabolic pathways can radically alter biological elimination rates and, therefore, by definition, biological half-lives. The biological half-life for radium is approximately 50 years.

3.10.3 EFFECTIVE HALF-LIFE

In order to determine how much of a radioactive substance will remain in the human body at a given time, another half-life term is defined which combines the effects of both the radiological and biological half-lives and is shorter than either. It is called the effective half-life ($T_{1/2eff}$).

The effective half-life of radium-226 is virtually the same as the biological half-life because the former is so much longer than the latter. For radium-228, the radioactive half-life is short enough to set the rate of its elimination.

3.10.4 SUMMARY

- The radioactive half-life of a particular radionuclide is the time taken for its activity to reduce by half due to radioactive decay.

- The biological half-life of a particular radionuclide is the time taken for half of it to be eliminated from the human body by natural biological means only.

- The effective half-life of a particular radionuclide is the time taken for the radioactivity of a particular radionuclide within the human body to decrease by half due to a combination of radioactive decay and biological elimination.

4

BIOLOGICAL AND HEALTH EFFECTS OF RADIATION

STUDY OBJECTIVES

This chapter will enable the student to:

- Define ionization.

- Describe the mechanisms of ionization by gamma, alpha, and beta radiation.

- Understand how ionizing radiation can cause biological damage.

- List and discuss the three factors which affect the sensitivity of a cell to radiation.

- Understand the difference between acute and chronic radiation exposure.

- Describe the possible biological effects of acute and chronic exposure.

- Name and explain the three basic models for dose response.

4.1 INTRODUCTION

Research on the biological and health effects of radiation has been going on for over 80 years. More data are available on the biological effects of radiation than on any other environmental health hazard to which we are exposed.

Radiation exposure control regulations have led almost all other regulations governing exposures of workers and the general public to potential chemical and toxic hazards. Partly due to the fearful attitude expressed toward radiation, standards for radiation protection are very strict. It is widely believed that exposure to radiation at levels below regulated limits is safe. The Nuclear Regulatory Commission (NRC), the governing body responsible for specifying national radiation protection standards, has stated that:

> Control of exposure to radiation is based on the assumption that any exposure, no matter how small, involves some risk. The occupational exposure limits are set so low, however, that medical evidence gathered over the past 50 years indicates no clinically observable injuries to individuals due to radiation exposures when the established radiation limits are not exceeded.

Radiation exposures to humans are subdivided into external doses (due to radiation from external sources) and internal doses (due to radionuclides deposited within the body). Internal doses are usually assessed as lifetime committed doses, i.e., the total dose that will result from the presence of the radionuclide in the body during the entire lifetime of the individual concerned. (For regulatory purposes, calculations are based upon a "lifetime" of 50 years.)

A large committed dose can result from a relatively small amount of radioactive material present in the body. This is in comparison to the dose that would result from short-term exposure to penetrating radiation from external sources. This remains true even when the elimination of radioactive material from the body (effective half-life) is accounted for. For predominantly alpha- and beta-emitting sources, the biological damage associated with external radiation exposure is minimal, whereas the damage that can be caused by such radionuclides inside the body is quite substantial. The amount of radioactive material present in a person's tissues is termed a body burden. Figure 4.1 shows how total radiation doses would accumulate over a week in a theoretical radiation worker who had cumulative

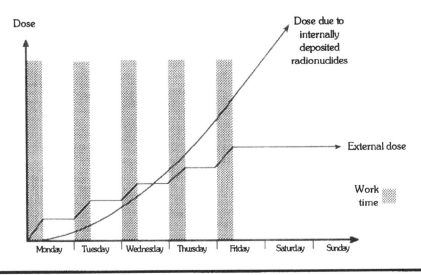

FIGURE 4.1   INTERNAL VS. EXTERNAL RADIATION EXPOSURE

external radiation exposure every day during a one-week period versus persons who acquired a small body burden each day in the same period.

As NORM radionuclides emit alpha and beta radiation, control of radioactive material and prevention of its assimilation into the body are often considered the most important aspects of radiation protection from NORM. The gamma radiation from NORM constitutes a much lesser risk.

Ingestion is the easiest pathway for radioactive material to enter the body. This is why it is imperative that no eating, drinking, smoking, or chewing take place in any area where NORM-related work or storage of NORM-contaminated material takes place. These and other simple practices such as washing the hands after handling potentially contaminated equipment are the most effective defense against the ingestion of radioactive material.

When vapors or aerosols containing radionuclides are present in the air, they can be deposited in nasal cavities and the lungs. From there, they can be absorbed or ingested. Many operations involving NORM, including soil moving, maintenance of contaminated equipment, and various decontamination methods, have a substantial chance of generating dust or spray. Control of airborne radioactivity is therefore an extremely important aspect of radiation protection. Methods of controlling exposure to airborne radionuclides are discussed in Chapter 8.

NORM can be ingrained into the skin, although this is very unusual. Open wounds are a significant concern but are not generally as important as inhalation or ingestion.

Although not a method for radioactive materials to enter the body, radionuclides can be formed inside the body by exposure to neutron radiation. This process is known as neutron activation. Neutron radiation is encountered in the oil industry in the form of radioactive sources used for well logging. However, activation of biological tissues is not a significant aspect of the hazard from these sources.

4.2 IONIZATION

Alpha, beta, and gamma radiation are known as ionizing radiation because of their ability to cause ionization. One alpha or beta particle or gamma ray will typically be able to ionize more than one atom. Ionization of atoms or molecules within body tissues is the mechanism by which radiation damages the body.

Atoms and molecules in nature generally exist in an electrically neutral state. When ionizing radiation interacts with atoms or molecules, it can remove electrons from them or it can break up molecules into separate parts with unbalanced charges. Electrically charged atoms or molecules are called ions. Many such ions are extremely chemically reactive and may react rapidly with any nearby chemicals, producing more complex ionic species. The chemically reactive ion pairs formed by radiation fall within the class of reactive chemicals known as free radicals.

Gamma radiation causes ionization by imparting sufficient energy to an electron to cause it to escape from its associated nucleus (as illustrated in Figure 4.2). This will result in the existence of a positively charged atom and a free electron, both of which are chemically reactive. In a similar fashion, gamma radiation can also impart energy to a chemical bond and break it, leaving two reactive atoms. This is a purely random event that requires the gamma ray to exactly strike an electron, which is why gamma radiation can penetrate so effectively.

A beta particle can impart kinetic energy to the electrons surrounding an atom by striking them or coming very close and exerting a coulomb repulsive force on them, as shown in Figure 4.3. Because of this additional mechanism of ionization compared to gamma rays, beta radiation will cause more ionization in a shorter distance. This is why beta particles cannot penetrate as effectively as gamma rays.

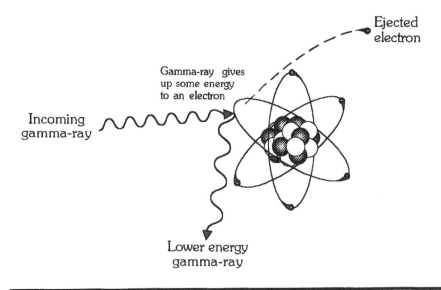

Ejected electron

Gamma-ray gives up some energy to an electron

Incoming gamma-ray

Lower energy gamma-ray

FIGURE 4.2   MECHANISM OF IONIZATION BY GAMMA RADIATION

Alpha particles have an even more efficient mechanism for causing ionization. Because of their double positive charge, they are highly attractive to electrons. An alpha particle will thus "strip" electrons from atoms that it passes closely (illustrated in Figure 4.4). Add to this an alpha particle's greater mass and kinetic energy, and it becomes a far more efficient ionizer

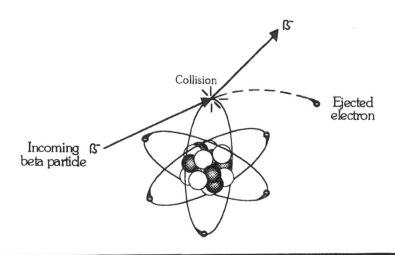

β⁻

Collision

Ejected electron

Incoming β⁻ beta particle

FIGURE 4.3   IONIZATION BY BETA RADIATION

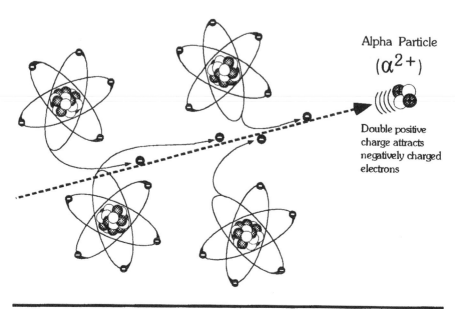

Alpha Particle

(α^{2+})

Double positive
charge attracts
negatively charged
electrons

FIGURE 4.4 Ionization by Alpha Radiation

than either beta particles or gamma rays. Because alpha radiation ionizes so effectively, an alpha particle will give up its energy in a short distance, which is the reason why alpha radiation has a poor penetrating ability.

4.3 The Effect of Ionizing Radiation on Body Tissue

There are two ways in which ionizing radiation can cause damage to cellular components or biochemicals. First, a component of a cell can be directly ionized by radiation. It will therefore experience some form of direct damage, although it may or may not be significant. Second, free radicals and other chemicals generated by ionizing radiation can chemically interact with and hence damage cell structures. Thus, damage to body cells can be considered either direct (ionization of the cell component itself) or indirect (chemical attack by free radicals or other active chemical species).

Body cells are composed mostly of water (over 80% of the mass of the human body is water). Therefore, radiation that interacts with body tissues will most probably interact with water in some way. When radiation ionizes water, a number of different ions can result. However, most of these will rapidly recombine with other atoms that may be present (again, most likely water) to form a variety of chemical compounds. In turn, these compounds will usually be chemically reactive. The ionic species and chemical compounds believed to be most commonly formed and most destructive are the hydroxyl radical (OH^- or OH^+) and hydrogen peroxide (H_2O_2).

Production of one free radical by another will continue until the reaction is terminated by reaction with a terminating agent. There are many such chemicals. They are a natural defense mechanism against unwanted, reactive compounds. Examples include the antioxidant vitamins A, C, and E and also so-called conjugating compounds such as glutathione and cysteine.

Most of the time, with low levels of radiation damage, any effect on the structure of a cell is minimal. Most of the major components of a cell are huge on an atomic scale, and a little radiation-induced damage will go almost unnoticed. Damage to large-scale cellular components and macromolecules on a substantial enough scale to cause medically significant cell damage requires enormous doses of radiation which can only be easily acquired when working with massive sealed sources, reactors, or powerful radiation-generating machines. Such devices are usually only encountered in some nuclear physics laboratories or where medical or industrial radiography is taking place. Even chronic exposure to moderately large doses of radiation does not result in severe damage to the overall structure of cells and tissues.

There is one component or class of components in a cell that can receive medically significant damage from relatively small doses of radiation. These components are the carriers of genetic information: ribonucleic acid (RNA) and deoxyribonucleic acid (DNA). The latter is particularly significant, as it contains the encoded information that determines and regulates virtually all bodily activity, including cell growth and division.

DNA consists of two chains of alternating sugar (deoxyribose) and phosphate groups. The sugars each carry one of four types of nitrogenous bases: cytosine, thymine, adenine, and guanine. The phosphate/sugar/base combinations are known as nucleotides (not to be confused with nuclides) and are linked together in a very specific order by special chemical bonds. The chains wind around each other in a spiral structure known as a double helix and are bonded to each other. Each nucleotide is the opposite of its complementary nucleotide on the other chain. Cytosine and guanine are always paired, as are thymine and adenine. The exact order of the four components defines everything about the structure and function of the cell and indeed those of every cell in the entire body. Each cell contains many strands of DNA, and a single strand of DNA will typically contain billions of nucleotide pairs. DNA is reproduced to pass on genetic information when a cell divides.

Each strand of DNA can be subdivided into recognizable units called genes. Each gene specifies a particular function or attribute of the cell or organism from which it came. If a particular gene becomes damaged, then the attribute or function controlled by that particular gene will also be affected.

In most cases, especially if the damage is not too severe, natural cell mechanisms are able to repair most DNA damage. Even when such damage goes unrepaired, provided the particular gene damaged does not control an important aspect of the cell's function, no ill effects will result. Should the damage to a gene be severe in the case where the gene controls a critical function of the cell, then the cell may cease to carry out its intended purpose and lose the ability to grow, reproduce, or die.

Human body cells naturally wear out or grow old and cease to function efficiently, at which point natural regulating mechanisms cause the death or destruction of the cell. Some cells, such as those which comprise the nervous system, will last a lifetime, whereas others, such as red blood cells or gametes, die in just a few weeks and need to be replaced. It is in order to grow, replace worn out or dead cells, and reproduce that cell division takes place. Cell division is the process whereby a cell divides into two daughter cells.

The ability of a cell to repair radiation damage successfully depends partly upon whether or not the cell is about to go through cell division. If the cell divides before damage can be repaired, any damage that was done to the DNA will probably be passed on to one or both of the daughters. As a cell prepares itself for and goes through division, chemical changes take place which make it possible for DNA to replicate itself. Unfortunately, these changes also make the cell less able to effectively deal with DNA damage. If damage has occurred to the DNA of a cell, there are four possible alternatives for the cell:

- The cell may repair the damage and be able to undergo normal cell division.

- Due to alteration of the DNA, the cell may produce abnormal daughters upon division. These daughters may possibly be cancerous.

- Due to alteration of the DNA, the cell may produce abnormal daughters that are incapable of surviving and die.

- The cell may die.

Provided too many cells in a particular part of the body do not suffer irreversible radiation damage or death, as described above, there should be no significant medical problem. Because the human body has many billions of cells, a few non-performing or dead ones will not even be noticed. However, there is one particular type of genetic damage that can be very serious. Should radiation damage occur to a specific type of gene called a proto-oncogene (which controls cell replication) and thus result in its conversion to an oncogene (which is a gene that allows uncontrolled cell replication), then a cancer cell could result. Even in such an instance, the development of a tumor is not necessarily the outcome because the human body has effective mechanisms to detect and destroy such abnormal cells.

Cells that have had their genetic material (DNA) altered slightly may be capable of continued life and growth. These alterations cause changes in the characteristics of the cell. These changes are called mutations and will be passed on to the daughter cells through their DNA. Popular literature will have the uninformed reader believe that such mutations can be passed on to future generations and could result in gross birth defects and other abnormalities. This situation has been seen to a certain extent in areas such as Hiroshima or in the vicinity of Chernobyl. However, statistically significant increases in rates of birth defects or other reproductive dysfunction attributable to radiation exposure to the parents are not seen unless such exposure has been severe enough to cause clinically observable effects (radiation sickness) in the parents.

4.4 RADIOSENSITIVITY

Some cells are much more sensitive to radiation damage than are others. Three major factors are believed to affect the radiosensitivity of a cell:

- The rate of cell division.

- The stage of cell division.

- The degree of cell specialization or differentiation.

It has been determined that the rate of cell division has the greatest effect on cell radiosensitivity. The faster cells in a particular tissue divide, the more sensitive they are to radiation. The stage of cell division is related to the rate of cell division, as rapidly dividing cells are more likely to be dividing or about to divide at any given time.

Cells which, by their nature or the stage of development of the organism that they constitute, undergo more rapid division are considerably more susceptible to radiation damage than cells which do not fit such a profile. It is for these reasons that a developing embryo or fetus is especially susceptible to radiation exposure. Children and young adults are also more susceptible to radiation than older adults. Regulations reflect this situation by severely limiting maximum permissible radiation doses to pregnant women and children under the age of 18.

The amount of specialization of a cell also affects it sensitivity to radiation. Cells that are undifferentiated and have no defined functions are unable to withstand such high doses of radiation as more highly differentiated cells that have specialized functions. Nerve cells, which serve only one purpose, are very resistant to radiation damage. Immature red blood cells (erythroblasts) in the bone marrow are very sensitive to radiation because they can perform the functions of many cells. Other cells that reside in the bone marrow, called stem cells, are responsible for producing most of the many different types of white blood cells that enable the body to fight infections and destroy cancerous cells. These stem cells are very susceptible to radiation, which causes a double problem when they are damaged or killed because the body is unable to deal with normally encountered pathogens and any cancerous cells that may arise as a result of the radiation exposure.

4.5 THE BIOLOGICAL EFFECTS OF SINGLE LARGE EXPOSURES TO IONIZING RADIATION (ACUTE EFFECTS)

Large doses of radiation can cause gross biological damage to the body and in some instances even death. However, the body can tolerate fairly high doses of radiation without experiencing adverse effects. A dose of 25,000 to 50,000 millirem (25 to 50 rem, 250 to 500 millisieverts) is required before there will be any medical evidence of the exposure. This evidence consists of mostly minor changes to blood chemistry and white blood cell count. Exposure totaling 100,000 millirem (100 rem, 15 sieverts) or more is required before radiation sickness is evidenced in most individuals. Symptoms of radiation poisoning include headache, fever, body aches, nausea, and diarrhea. At increasing doses above this level, the spectrum and severity of symptoms increase. Eventually, intellectual impairment, tremors, and coma would result, usually preceding death. Depending upon the general health of the individual and the availability of medical assistance, doses in excess of 500 to 1000 rem (5 to 10 sieverts) usually prove to be fatal. The likely effects of various, very large, and acute doses of penetrating radiation are listed in Table 4.1.

There is no recorded instance of male infertility due to radiation exposure without the male in question having experienced symptoms of severe radiation poisoning. It is believed that a whole-body dose sufficient to result in irreversible sterility would quite likely prove fatal.

The medical symptoms of radiation sickness are due to three main effects:

- Saturation of natural mechanisms for dealing with free radicals results in a build-up of toxins (including radiation-induced free radicals) which would normally be cleared from the body.

- Radiation-damaged cells cease reproduction while they are repairing the damage. When this occurs to cells that normally divide very frequently, such as the bone marrow cells that produce blood cells, blood cell counts will decrease until the cells have recovered. Crypt cells in the lining of the gut which produce cells to replace those worn off by the passage of food through the digestive tract are also susceptible to radiation damage.

- Radiation damage to cell membranes can make them "porous." This results in a great loss of body fluids with accompanying dehydration. Damage to the membranes surrounding specialized internal cellular components, called lysozomes, results in the internal release of digestive enzymes which can destroy many other internal molecules. Similar radiation-induced impairment of the insulating sheath surrounding nerve cell axons and the membranes across which electrochemical signals pass results in the symptoms characteristic of nerve damage, seen in cases of extremely high exposures.

TABLE 4.1 SUMMARY OF THE CLINICAL EFFECTS OF ACUTE DOSES OF PENETRATING RADIATION (FROM U.S. GOVERNMENT STUDY ON THE EFFECTS OF NUCLEAR WEAPONS)

Range	0–100 rem (0–1 Sv)	100–200 rem (1–2 Sv)	200–600 rem (2–6 Sv)	600–1000 rem (6–10 Sv)	1000–5000 rem (10–50 Sv)	Over 5000 rem (50 Sv)
Incidence of vomiting	None	100 rem: 5% 200 rem: 50%	300 rem: 100%	100%	100%	
Delay time	—	3 hours	2 hours	1 hour	30 minutes	
Leading organ	None	Blood forming tissues			Gastrointestinal tract	Central nervous system
Characteristic signs	None	Decreased number of white blood cells	Severely decreased white blood cells, hemorrhage, infection; hair loss above 300 rem		Diarrhea, fever	Convulsions, tremors, incoordination
Critical period post-exposure	—	—	4–6 weeks		5–14 days	1–48 hours
Therapy	Reassurance	Reassurance, blood monitoring	Blood transfusion, antibiotics	Possibly bone-marrow transplantation	Maintain electrolyte balance	Sedatives
Prognosis	Excellent	Excellent	Good	Guarded	Hopeless	
Convalescent period	None	Several weeks	1–12 months	Long	—	—
Incidence of death	None	None	0–80% (variable)	80–100% (variable)	Almost 100%	
Death occurs within	—	—	2 months		2 weeks	2 days
Cause of death	—	—	Hemorrhage, infection		Circulatory collapse	Respiratory failure, brain edema

4.6 The Biological Effects of Continual or Repeated Exposure to Low Levels of Ionizing Radiation (Chronic Effects)

The majority of NORM workers will not experience acute radiation exposures but will acquire very small doses every day that they work with or around NORM. Typically, this will take place frequently over many years. The levels of whole-body radiation exposure are usually low in comparison to background levels, and the potential for receiving extremely large, acute doses of radiation is usually negligible.

While much epidemiological information has been gathered about the effects of large acute radiation doses, very little is known about the commonly encountered small, chronic exposures. Most of the data available are based upon statistical evidence and extrapolations of data from acute exposures and radiation medical treatments. For very low doses, the effects are virtually indistinguishable from background. However, in accordance with the approach taken for toxic chemicals, various agencies and the federal government have taken the position that any exposure to radiation could be potentially harmful.

The basis for control of radiation exposure is the assumption that any exposure, no matter how small, involves some risk. However, this risk is small when compared to the risks of everyday life if this exposure is kept within acceptable guidelines. In 1954, the National Council on Radiation Protection and Measurements (NCRP) issued a report in which the phrase "acceptable risk" was described:

> As a matter of principle, it is sound to avoid all unnecessary exposure to ionizing radiation because it is desirable not to depart from the natural conditions under which man has developed by evolutionary processes. However, man has always lived in a field of ionizing radiation due to the presence of radioactive material in the earth and to cosmic rays. Whether exposure to this level of radiation is beneficial or deleterious to man (and the race) is a matter of speculation. The obvious fact is that it cannot be avoided and it is, therefore, normal for man to live in this environment. We have then a lower limit of continuous exposure to radiation that is (unavoidably) tolerated by man.

The report also stated:

The only statement that can be made at the present time about the lifetime exposure of persons to penetrating radiation at a permissible level considerably higher than the background radiation level, but within the range of radiological experience, is that appreciable injury manifestable in the lifetime of the individual is extremely unlikely. It is, therefore, necessary to assume that any practical limit of exposure that may be set up today will involve some risk of possible harm. The problem then is to make this risk so small that it is readily acceptable to the average individual; that is, to make the risk essentially the same as is present in ordinary occupations not involving exposure to radiation.

In issuing its first recommendations regarding radiation exposure in 1958, the International Commission on Radiation Protection (ICRP) stated:

...the Commission recommends that all doses be kept as low as practicable and that any unnecessary exposure be avoided.

It then went on to say:

Any departure from the environmental conditions in which man has evolved may entail a risk of deleterious effects. It is therefore assumed that long continued exposure to ionizing radiation additional to that due to natural radiation involves some risk. However, man cannot entirely dispense with the use of ionizing radiation, and therefore the problem in practice is to limit the radiation dose to that which involves a risk that is not unacceptable to the individual and the population at large.

4.7 RADIATION DOSE/RESPONSE MODELS

Dose/response models provide a useful way to examine radiation risk. There has been much discussion and argument over whether deleterious effects of radiation exposure require a minimum or threshold dose or whether very small doses produce very small effects. Two common models are used for describing dose/response: the linear dose/response model and the threshold model. A third, somewhat controversial, model suggests that small doses of ionizing radiation are actually beneficial. This is known as the

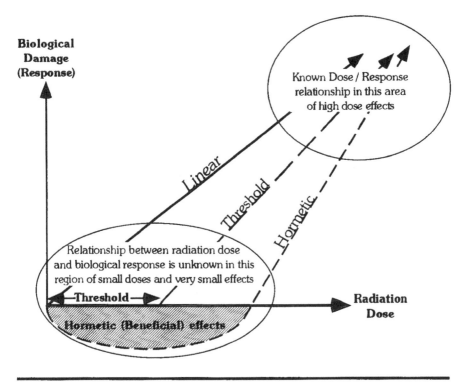

FIGURE 4.5   RADIATION DOSE/RESPONSE RELATIONSHIPS

theory of radiation hormesis. The three dose/response curves are shown in Figure 4.5.

4.7.1 LINEAR DOSE/RESPONSE

The linear dose/response model assumes that the relationship between the dose received and the effects of that dose is simple and linear. This implies that even exceptionally small doses of radiation can cause some type of biological response. Man and the organisms from which he evolved have always been exposed to low "background" levels of radiation. As a result of this, many scientists doubt the validity of the linear model. Research has shown that the biological damage caused by exposure to very high levels of radiation does exhibit linear qualities, as depicted in Figure 4.5. Because it is conservative in nature, the linear dose/response model forms the basis for current radiation protection practices and procedures.

4.7.2 THRESHOLD DOSE/RESPONSE

This model assumes that a certain cumulative dose must be reached before the effects of any radiation damage will be manifested. The model contains some merit because all life on earth is constantly exposed to radiation from natural sources and should have evolved to deal with it. It is certainly true that at very low radiation exposure, the effects of that exposure are statistically insignificant, but this does not mean that there is no effect.

Recently, a new development in theories of radiation exposure has received much attention. It centers on the concept of radiation hormesis and "radiation deficiency." In an article published in 1982, Professor Luckey of the University of Missouri, Columbia, stated, "Extensive literature indicates that minute doses of ionizing radiation benefit animal growth and development, fecundity (ability to produce offspring), health and longevity. Specific improvements appear in neurological function, growth rate and survival of young, wound healing, immune competence, and resistance to infection, radiation morbidity (radiation sickness), and tumor induction and growth."

Many environmental chemicals and medical drugs exhibit hormetic behavior, i.e., they are beneficial at low doses but harmful at high doses. The theory of radiation hormesis is similar. It holds that at low doses, radiation exposure may actually be medically beneficial despite the fact that higher doses are known to cause biological damage. The beneficial effects of exposure to low levels of radiation are believed to include:

- Increased life span.
- Increased growth and fertility.
- Reduction in incidence of cancer.

In association with radiation hormesis is the concept of radiation deficiency, which refers to the possibility that biological systems may actually need radiation, somewhat like vitamins or certain trace elements, in order to maintain health. This suggests that an organism would show signs of ill health if it were not allowed to receive the required amount of beneficial radiation. This may be because some of the free radicals formed within body cells are useful to our bodies and not easily formed by other means.

The theories of radiation hormesis and radiation deficiency have generated significant discussion, but studies that have been performed have often arrived at conflicting conclusions.

5

RADIATION IN THE WORLD AROUND US

STUDY OBJECTIVES

This chapter will enable the student to:

- List and describe natural sources of radiation exposure.
- List and describe artificial sources of radiation exposure.
- Name the main radionuclides that contribute to external terrestrial radiation exposure.
- Name the main radionuclides that contribute to internal terrestrial radiation exposure.
- List some common consumer products that contain radioactive material.
- Identify the artificial source of radiation that accounts for the greatest dose to the average person in the United States.
- List common occupational sources of radiation exposure.
- Contrast occupational radiation exposures due to NORM with those from other sources.

5.1 INTRODUCTION

Every organism on earth is exposed to radiation for every minute of every day of its entire life. Mankind is no exception.

In addition to the radiation exposure that workers may receive as a result of the job they perform, they will also be exposed to radiation from the

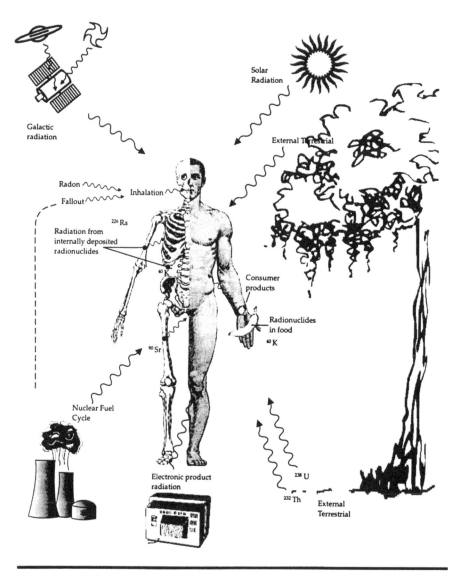

FIGURE 5.1   SOURCES OF RADIATION TO WHICH THE AVERAGE HUMAN IS EXPOSED

environment in which they live. All individuals are subject to irradiation even though they may not work with radioactive materials. For oil field workers, who work with or around NORM, the dose that they receive from sources other than NORM is probably far more significant than the dose due to NORM.

The total radiation to which an average person is exposed can be broken down into two major categories, based on whether its origin is natural or artificial (i.e., man-made). Natural sources include cosmic radiation and external and internal terrestrial (i.e., "of the earth") radiation. Artificial sources consist of medical and dental x-rays, fallout, consumer product radiation, and nuclear power generation. Occupational radiation exposures also fall into this category.

A diagrammatic representation of all sources of radiation to which the average human is exposed is presented in Figure 5.1. A breakdown of the typical radiation exposure from all sources for the average person living in North America is represented in Figure 5.2. The dose due to cosmic radiation is based on exposure at sea level. (For persons living in Denver, Colorado, for example, this dose contribution would be 1.5 to 2 times as large as shown.) The relative doses from soil and radon are based upon the national average. In certain geographic areas, these can be somewhat higher or lower. Doses due to medical oncology or other uncommon forms of high radiation exposure are not included because they apply to such a small fraction of the population.

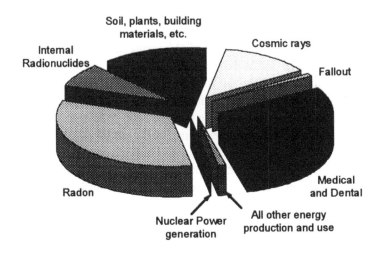

FIGURE 5.2   BREAKDOWN OF ANNUAL RADIATION DOSE TO THE AVERAGE U.S. ADULT

5.2 NATURAL SOURCES OF RADIATION

Natural sources of radiation include all those sources to which we are exposed that originate from sources that exist without any intervention (or control) from mankind. Such sources of radiation may be conveniently categorized as external terrestrial, internal terrestrial, and cosmic.

5.2.1 EXTERNAL TERRESTRIAL RADIATION

Many different naturally occurring radionuclides are present in the earth. They are found not only in geological formations, but are also present in the soil and in all organisms, including man. Most of the external dose to man from terrestrial sources is due to radon gas. Estimates of the dose to different tissues due to radon vary. Data from Sweden indicate that the lungs receive the highest exposure. However, radon exposure is considered to be external. Radon originates in the soil and in building materials. The radon isotope of primary concern is radon-222 (^{222}Rn), which is a daughter of uranium-238. Other radon isotopes are of less concern due to their very rapid decay.

Much of the remaining external terrestrial radiation dose is gamma radiation from the many radionuclides in the uranium-238 and thorium-232 decay chains and from potassium-40. In a few parts of the world, exceptionally high concentrations of these substances are found in soil and certain geological formations. The uranium-238 and thorium-232 decay chains are depicted in Figure 5.3. There is also a small external dose contribution due to rubidium-87 and radionuclides in the uranium-235 decay chain. All of these radionuclides are termed primordial because they have very long half-lives and have existed for as long as the earth itself.

Several studies have been performed in an effort to measure average dose rates to the world population due to naturally occurring external terrestrial radiation. Sufficient data have not yet been collected to make a firm estimate. The studies that have been completed contain data on both indoor and outdoor dose rates. Using these data and making assumptions about the time spent indoors, the estimated average dose rate from these sources worldwide, excluding radon, is approximately 0.5 millisieverts (50 millirem) per year.

In parts of India and Brazil, among other places, the annual external terrestrial dose exceeds 10 mSv (1 rem) per year. This is due to the presence of veins of a thorium mineral called monazite at the surface of the earth.

Uranium -238 Decay Chain

^{238}U   4.47x10^9 yrs
⇓ α
^{234}Th   24.1 days
⇓ β
^{234}Pa   6.7 hrs
⇓ β
^{234}U 2.45x10^6 yrs
⇓ α
^{230}Th 7.54x10^4 yrs
⇓ α
^{226}Ra   1608 yrs
⇓ α
^{222}Rn   3.82 days
⇓ α
^{218}Po   3.11 min
⇓ α
^{214}Pb   26.8 min
⇓ β
^{214}Bi   19.9 min
⇓ β
^{214}Po   164 μsec
⇓ α
^{210}Pb   22.3 yrs
⇓ β
^{210}Bi   5.01 days
⇓ β
^{210}Po 138.4 days
⇓ α
^{206}Pb   (Stable)

Thorium-232 Decay Chain

^{232}Th 1.4x10^{10} yrs
⇓ α
^{228}Ra   5.75 yrs
⇓ β
^{228}Ac   6.13 hrs
⇓ β
^{228}Th   1.91 yrs
⇓ α
^{224}Ra   3.66 days
⇓ α
^{220}Rn   55.6 sec
⇓ α
^{216}Po   0.15 sec
⇓ α
^{212}Pb   10.64 hrs
⇓ β
^{212}Bi   60.6 min
64%   36%
β ↙   ↘ α
^{212}Po 0.3 μsec ^{208}Th 3.05 min
α ↘   ↙ β
^{208}Pb   (Stable)

FIGURE 5.3   URANIUM-238 AND THORIUM-232 DECAY CHAINS

The average U.S. adult exposure due to radon gas is roughly 2 mSv (200 mrem). The amount of radon in the air in any geographic area depends on the uranium and thorium concentrations in the soil. The weather will also greatly affect the concentration of these gases. It is common to find elevated levels indoors compared to outdoors. This is dependent upon building materials, construction, and the ventilation rate of a building.

There are many areas of the world where radon exposures are high. Such exposures seem to be mainly related to radon and radium dissolved in water supplies. High levels of dissolved radon and radium are most common in drinking water supplies that originate underground in areas with igneous rocks. This is especially true in areas with geothermal or volcanic activity. In areas such as caves and mines, particularly metalliferous mines, radon levels can be excessive. British cave researchers have discovered strange relationships between weather conditions and radon levels underground. Cold weather and high pressure seem to be linked to order-of-magnitude increases in radon levels in certain cave systems.

Sleeping with another person causes an individual to receive a calculated dose of between 0.10 and 3 mrem (1 to 30 μSv) per year due to gamma rays originating from radionuclides in the other person's body. The wide range of doses quoted is due to the strong dependence of the calculated value upon how close the two people sleep.

5.2.2 Internal Terrestrial Radiation

Most of the radionuclides found in the external world are transferred to man via the food chain. No study has yet been able to correlate the relationship between concentrations of naturally occurring radionuclides in soil and those found in human populations.

The main radionuclides found in the human body are potassium-40 (^{40}K), radium-226 (^{226}Ra), radium-228 (^{228}Ra), carbon-14 (^{14}C), tritium (^{3}H), and sodium-22 (^{22}Na). Of these, potassium is the most abundant, contributing approximately 80% of the total dose from all internal emitters. Potassium-40 is found in all organs in the body but its concentration varies greatly. Any of the foodstuffs known to contain potassium will contain potassium-40. Foods known to be especially rich in potassium include bananas and cantaloupe.

It has been estimated that the average body burden of radium is approximately 125 picocuries. Of this amount, 80% is believed to reside in the bones. This can be explained by the fact that radium has similar chemical properties to calcium, which is a key component of bone. A food containing unusually high quantities of radium is the Brazil nut.

TABLE 5.1   AVERAGE ANNUAL RADIATION DOSE TO THE AVERAGE ADULT IN THE U.S. DUE TO INTERNALLY DEPOSITED, NATURALLY OCCURRING RADIONUCLIDES

	Dose Rate (mrem/year)	
Source of Irradiation	Gonad	Bone Marrow
Potassium-40	20	15
Radium-226 and daughters	0.5	0.6
Radium-228 and daughters	0.8	1.0
Lead-210 and daughters	0.3	0.4
Carbon-14	0.7	1.6
Radon-222 (absorbed in bloodstream)	3.0	3.0
Total	**125**	**122**

The last three radionuclides identified are of particular interest in that they are naturally occurring but not primordial in origin. They are formed by the action of cosmic radiation on gases in the upper atmosphere. Carbon-14 can be used to date the death of an organism because once an organism has ceased to respire, carbon is no longer assimilated. The ratio of carbon-14 to its non-radioactive counterpart, carbon-12, will vary with time, as carbon-14 undergoes radioactive decay.

The dose rates to selected organs from the main contributing natural internal radionuclides are shown in Table 5.1.

5.2.3 COSMIC RADIATION

Cosmic radiation was discovered while experimenters were trying to find ways to reduce background radiation levels. Detection devices showed a response even though no sources of radiation were present. It was assumed that all this background radiation was coming from natural sources in the ground and building materials. The scientists surmised that if the detector were sufficiently elevated, the background should be reduced. Balloons were used to carry radiation detection instruments to great heights, but this yielded higher levels of radiation. This and other data proved that this radiation was actually coming from outer space; hence, it was termed "cosmic" radiation.

The origins of cosmic radiation can be categorized as either solar or galactic. In either case, the radiation originates in forms considerably more "exotic" than those seen on earth. Interaction of such radiation with the earth's atmosphere and magnetic field results in the familiar gamma ray background that is observed at the earth's surface.

Solar radiation is radiation that originates in the sun. The sun gives off a steady stream of radiation all the time; however, most of the dose due to solar radiation is a result of severe solar flares that occur periodically. Solar flares can be likened to storms on the surface of the sun. They consist of huge jets of super-heated material which is believed to have been ejected from deep within the sun. Protons constitute the majority of the radiation coming from solar flares. Solar flares are closely associated with sun-spot activity which occurs in 11-year cycles. The dose rates at high altitudes can reach hazardous levels during intense solar flare activity. The largest solar flare ever measured caused a dose rate of 10 rem per hour at an altitude of 80,000 feet. Because of the possibility of increased radiation levels due to solar flares, commercial passenger aircraft are required to have an operating radiation detector on board to alert the pilot to excessive radiation exposure.

Galactic radiation is composed mostly of charged particles and gamma rays and comes from outside the solar system. The particulate portion of galactic radiation consists of 87% protons, 12% alpha particles, and 1% other heavy nuclei. These particles may have energies that range from a few electron volts to greater than 10^{17} (100 quadrillion) electron volts (which is many millions of times higher than the energy of radiation commonly detected at the earth's surface). Galactic gamma ray radiation exhibits energies up to at least 3×10^{10} (30 billion) electron volts. The highest energies of gamma radiation could be much higher, but gamma ray detectors capable of reading higher energies have yet to be placed in space. The Compton orbiting gamma ray observatory has observed no drop off in the amount of gamma radiation at energies approaching the aforementioned limit; therefore, higher energy gamma rays are almost certainly present.

Galactic radiation actually includes extragalactic radiation from other galaxies and distant objects within the universe. Much of this radiation is at energies too high to have come from radioactive decay. Some astronomical objects such as quasars, pulsars, and supernovae produce radiation because of their intense gravitational or magnetic fields. They can produce ultra-high energy forms of radiation similar to those of which the abandoned superconducting supercollider (SSC) was supposed to have been capable. Indeed, the fantastic capabilities of the SSC would be totally insignificant compared to the conditions which most astrophysicists believe exist naturally in the vicinity of some black holes, for instance.

When cosmic radiation is measured at the surface of the earth, it is difficult to determine its origin. The radiation that is measured can be divided into two categories: primary and secondary.

Primary cosmic radiation is that which passes through the atmosphere and is detected directly. It consists mostly of very high-energy, low-mass particles and gamma rays. Interactions that occur as a result of these and heavier, more highly charged particles colliding with the atmosphere produce secondary cosmic rays. High-energy particles interact with atoms of the earth's atmosphere to produce electrons, photons, protons, neutrons, and other sub-atomic particles. These particles, in turn, produce further forms of radiation as they collide with elements in the atmosphere or decay on their way to the earth's surface. This multiplication may produce as many as 100 million secondary radiation events as a result of one primary event. At sea level, the average annual whole-body dose has been estimated at 30 mrem. It has also been estimated that approximately 20 mrem per year is the dose received from cosmic neutron radiation. This gives a total dose rate from cosmic radiation sources, at sea level, of approximately 50 mrem. This figure increases rapidly as altitude increases, as shown in Table 5.2.

Exact measurements of the increases in radiation exposure with altitude are somewhat difficult to make because any closed, pressurized, conducting vessel, such as an aircraft fuselage, will act to amplify the exposure levels.

TABLE 5.2 VARIATIONS IN ANNUAL DOSE DUE TO COSMIC RADIATION WITH ALTITUDE

Altitude (feet)	Dose Rate (mrem/year)	Example
Sea level	30	Gulf Coast
5,000	55	Denver, Colorado
10,000	140	Leadville, Colorado
30,000	1,900	Normal airplane
50,000	8,750	Concorde
80,000	12,200	Spy plane
120,000	10,500	Low-orbit spacecraft

Note: The doses shown above do not include the dose due to neutron radiation because the dose/altitude information relating to neutron dose is unavailable.

The dose due to cosmic radiation increases at higher latitudes (farther from the equator) because the earth's magnetic field tends to "funnel" charged particles toward the earth's magnetic poles. This is believed to be the cause of the Aurora Borealis (or Northern Lights), which only occurs toward polar latitudes. The combination of altitude and latitude means that passengers and crew typically receive higher doses in transpolar commercial air routes than in longer, lower latitude routes.

5.3 ARTIFICIAL SOURCES OF IONIZING RADIATION

5.3.1 FALLOUT

The term "fallout" refers to airborne radioactive material that falls to the earth following a nuclear bomb detonation or a similar atmospheric release of radioactive material. Such a release was the reactor explosion at Chernobyl, Ukraine on April 27, 1986. Radionuclides, identified as having originated from Chernobyl, were rapidly dispersed all around the northern hemisphere, including the northern United States. In many northern parts of Europe, including much of Britain, livestock, produce, and dairy products could not be consumed because of excessive radioactive content. Mankind has been exposed to some level of fallout since shortly after the first test explosion of a nuclear weapon on the Alamogordo air base in New Mexico. Sediments from the farthest reaches of both the Arctic and Antarctic poles contain distinctive radionuclides derived only from atomic weapons.

The intense heat of a nuclear explosion causes matter in the vicinity of the bomb to instantly vaporize. A mixture of radionuclides formed by the fission process, unused bomb fuel, and neutron-activated material originating from anything else nearby is caught in the fireball. As the rising cloud of superheated material reaches cold, low-pressure layers, high in the atmosphere, the characteristic mushroom-shaped cloud forms. The height to which the mushroom cloud rises is dependent on the yield (or explosive strength of the bomb), the mode of detonation (underground, surface, or air-burst), and the meteorological conditions in the area. For yields in the megaton range (releasing explosive energy equivalent to 1 million tons of the high explosive TNT), the cloud top may reach 25 miles in height.

In the 1950s, the United States had an active atmospheric test program for nuclear devices. The majority of tests were carried out at the Nevada Test Site. In 1963, the Limited Test Ban Treaty was signed. This forbade the surface testing of nuclear devices. All U.S. tests since then have been performed underground. Tests at the Bikini atoll in the South Pacific were conducted to ascertain the effect of atomic weapons on ships and other forces. Atomic bombs were detonated in the air at various altitudes above a "fleet" of scrap ships. Planes made daring flights into the mushroom clouds to measure radiation levels and determine whether military operations could continue in the aftermath of an atomic explosion.

The current exposure from fallout is estimated to be approximately 1 mrem (10 μSv) annually. However, older adults in particular have measurable amounts of radioactive cesium and strontium isotopes in their bones, from which they receive a small, continuing exposure.

5.3.2 CONSUMER PRODUCT RADIATION

Various consumer and industrial products contribute to the dose the public receives because they contain radioactive materials. A small number of people still wear luminous dial wristwatches that contain radioactive material. The radioisotopes that are primarily used are radium-226 and tritium (^{3}H). Radium-226 emits gamma radiation and, therefore, produces a more significant dose to the watch wearer than does tritium, which is a pure beta emitter and thus causes a very small level of secondary bremsstrahlung radiation.

Tobacco contains significant levels of lead-210 (^{210}Pb) and polonium-210 (^{210}Po), both alpha emitters. Approximately 50 million people in the United States smoke 20 to 30 cigarettes per day. These individuals receive an estimated annual dose to their lungs of approximately 8000 mrem (80 mSv). This exposure is from radionuclides taken up by the tobacco plant. Many plants are known to absorb lead from the soil and will therefore build up radioactive lead-210 in their tissues.

Radionuclides are also released into the environment from the combustion of fossil fuels. Coal burned to generate electrical power creates fly ash with measurable levels of radium-226, thorium-232, and isotopes of uranium, lead, and polonium. Radiation doses to the public from the burning of coal in the United States average 4 mrem per year. Dose rates to the bone could reach or exceed 36 mrem per year for those who live 500 yards downwind of a 1000-megawatt plant. Natural gas is also a source of radiation; it contains 10 to 20 picocuries of radon gas per liter. Natural gas appliances used in homes can cause a dose to the lungs of from 6 to 9 mrem (60 to 90 μSv) per year. More examples of radiation exposure from consumer products are given in Table 5.3.

TABLE 5.3 AVERAGE ANNUAL DOSE FROM CONSUMER PRODUCTS

	millirem	microsieverts
Radium dial alarm clock	7–9	70–90
Building masonry	7	70
Road construction materials	4	40
Home ionization smoke detector	1	10
Dental porcelain (to gum)	60,000	600,000
Rose-tinted glasses (to eye)	4,000	40,000
Aircraft luminous instrument dial	1,000–5,000	10,000–50,000
Radioactive lightning rod	0.05	0.5
Uranium-glazed dinnerware (to skin)	2,400	24,000
Gas camping lantern mantle	0.1–0.41	1–4

5.3.3 MEDICAL AND DENTAL X-RAYS

Most individuals have had a medical or dental examination involving an x-ray at one time or another. X-rays are used for a wide variety of diagnostic purposes. Typical doses associated with some of the more common diagnostic x-ray techniques are as follows:

- The typical chest x-ray gives the patient a dose of 80 to 100 mrem (800 to 1000 μSv).

- The typical dose to the jaw during a full set of dental x-rays can be up to 900 mrem (9 mSv). The whole-body dose from dental x-rays can be maintained below approximately 3 mrem (30 μSv) when modern dental x-ray equipment and techniques are utilized.

- A fluoroscope is a type of diagnostic x-ray device that is occasionally used to investigate gastrointestinal disorders. The typical dose rate received from such an instrument is 5 rem (5000 mrem, 50 mSv) per minute. In the 1940s and 1950s, similar devices were used to assist in fitting shoes; not surprisingly, they have long since been removed from shoe stores.

It should be understood that the above-mentioned doses are typical of equipment that is currently in use. The most modern equipment may yield lower doses, while older equipment could give rise to significantly higher doses.

Medical and dental x-rays are the greatest single category of contribution to the general public's radiation exposure each year. They account for an average of approximately 75 mrem (750 µSv) per person annually. This does not include therapeutic radiation treatments for cancer and other non-operable ailments, which may deliver doses in excess of several hundred rem to individual parts of the body. Such exposures are of an isolated nature and do not affect a large percentage of the general public.

5.3.4 NUCLEAR POWER GENERATION

The radiation exposure that a worker in a nuclear power plant receives due to power generation is primarily in the form of neutron and gamma radiation or from contamination that exists in plant equipment and components. Since 1973, the Nuclear Regulatory Commission (NRC) has compiled a report of the annual doses to workers in the plant who were actually monitored for radiation and who received a measured, non-zero dose. The dose currently averages less than 700 mrem (7 mSv) per worker-year, which is down from a high of 940 mrem (9.4 mSv) per worker-year. Changes to procedures and technology should result in this figure falling steadily throughout the coming years.

The NRC is required by federal law to determine the dose annually to persons living in the vicinity of nuclear power generating facilities. These doses are calculated using the actual amounts of radioactivity discharged by the power plant into the air and water. The typical annual dose to persons living within a 50-mile radius of a U.S. commercial nuclear power plant is less than one-hundredth of a millirem (0.1 µSv). This is 100 to 400 times lower than the dose received by someone living next to a coal-burning power plant. It is equivalent to receiving an extra 40 minutes of natural background radiation per year or to the increase in cosmic radiation exposure that would result from living at an elevation some 16 inches higher.

5.4 OCCUPATIONAL RADIATION EXPOSURES DUE TO NORM

In general, the radiation exposures that workers and the public receive due to NORM are very small in relation to those from other occupational and non-occupational sources. However, the regulatory standpoint is that any radiation exposure may have a small risk associated with it. Therefore, it is necessary to measure and control radiation exposures due to NORM.

Conventional means of monitoring radiation exposures and uptake of radioactive material are mainly intended to be used for exposures at significant proportions of allowable limits. In many cases, exposures due to NORM are very small and therefore difficult to quantify. There are, however, conceivable scenarios where hazardous or otherwise significant exposures could be obtained. This is the reason why NORM is regulated.

To appreciate the magnitude of radiation exposures that most workers are likely to receive from NORM, it is worthwhile to examine a few examples. As an extreme example of external radiation exposure due to NORM, consider the following.

A worker in a pipe yard works every day cleaning production tubing, much of which is contaminated with NORM. Radiation controls have not been implemented and pipe scale is allowed to pile up at the end of the cleaning machine. The worker spends the entire working day at the end of the machine, straddling a pile of pipe scale. The scale could conceivably have an activity of several thousand picocuries per gram. This could result in a radiation exposure rate at a distance of 18 inches of up to, say, 3000 microroentgen per hour. Using an approximate quality factor of 1 for the penetrating gamma radiation, the worker would receive a deep equivalent dose to his legs and gonads of 3 mrem (30 μSv) per hour. Over a complete year, if the worker were to put in 2000 hours, he could potentially receive a total dose of 6 rem (60 mSv). This would slightly exceed the maximum allowable dose of 5 rem (50 mSv) annually to his upper legs and gonads. Of course, even if the dose were to be slightly below 5 rem (50 mSv), this situation would be unacceptable because the ALARA principle would not be being observed. Simple controls could significantly reduce the dose received by workers in such a situation.

Federal law now requires that internal and external doses be assessed collectively. For the purposes of demonstration, we will consider only internal doses due to NORM in a similar scenario to that already outlined. Imagine that the same worker is using no respiratory protection and on an average day inhales 1 gram of pipe scale dust. (This figure is chosen for the sake of example only and is probably many times higher than that which could actually occur.) The worker may also ingest a small amount of dust, due to various factors such as inappropriate behavior while working with NORM (smoking, drinking, or chewing) and poor hygiene practices (hand and face washing). This would probably be insignificant compared to the inhalation, and its effects will therefore be disregarded. Similarly, the effects of radon inhalation would be less than those of inhaling NORM dust directly. The calculations for committed dose equivalent due to such an uptake are complex and dependent on many factors. However, if the NORM scale had an activity of 2000 picocuries (74 becquerels) per gram, the worker's annual intake would amount to a total of 500 nanocuries (18.5 kilobecquerels), which is just below the regulatory limit of 600 nanocuries (22.2 kilobecquerels).

Combining the effect of the two scenarios outlined, the worker in question would receive a total committed dose each year of approximately 10 rem (100 mSv), which is double the allowed limit. However, as can be seen from the assumptions being made, this would be a very extreme case. It is doubtful whether any oil field workers would ever actually work in a situation such as this, although some may have come close in the past.

Let's examine a more realistic situation in which a worker is directly handling and working around NORM-contaminated equipment for less than 2 hours per day on average. The worker encounters radiation exposure rates that do not exceed 1000 microroentgen per hour and average only, say, 100 microroentgen per hour. This worker takes appropriate precautions such as not eating, drinking, smoking, or chewing when working around NORM. He washes his hands and face before eating or leaving for home. The work does not require respiratory protection, but the worker inhales 10 grams of NORM dust per year, with an average activity of 200 picocuries (7.4 becquerels) per gram. The total committed dose equivalent that this worker would acquire per year would be around 65 mrem (650 μSv). This is just over one-hundredth of the allowed maximum and would add less than 50% to the worker's exposure due to background radiation. Such a dose would be less than that obtained from a single chest x-ray.

The figure calculated above would apply to a worker who had considerably more exposure than the average person who works in a NORM-contaminated oil field on a day-to-day basis. Experience has shown that most workers who are monitored by wearing dosimetry monitoring devices while at work receive no measurable dose that is distinguishable from normal background. This would normally mean that they received less than 5 mrem (50 μSv) annually in excess of the dose that they would have received without working around NORM.

While the internal and external doses derived for the scenarios previously outlined were roughly equivalent, it is generally held that the exposure problem associated with NORM is of an internal nature. To claim that external exposure due to NORM is not significant would be incorrect.

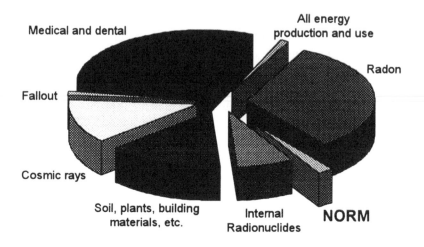

FIGURE 5.4   BREAKDOWN OF ANNUAL RADIATION DOSE TO THE AVERAGE OIL FIELD WORKER RECEIVING 5 MILLIREM PER YEAR OCCUPATIONAL EXPOSURE FROM NORM

However, gamma radiation from NORM should mainly be considered to be indicative of the presence of alpha-emitting contamination rather than of major concern in itself. The nature of radium is such that its main hazardous properties are due to internally deposited alpha- and beta-emitting radionuclides. There are no proven documented cases of any disease being directly related to exposure to NORM. The main risk factor from long-term exposure to NORM is probably a very small increase in the chance of contracting radiation-induced diseases, particularly those known to be related to radium, such as bone cancer or certain types of anemia. A breakdown of radiation exposures to a worker who receives a 5-mrem (50-μSv) annual occupational dose from NORM is illustrated in Figure 5.4.

NORM in the oil and gas industry is regulated to minimize the internal exposure of personnel by preventing or minimizing the inhalation or ingestion of alpha-emitting contamination. In other words, the external whole-body radiation received from NORM is usually not significant compared to radiation received from other unavoidable sources. However, NORM is of radiological concern due to the possibility of inhaling or ingesting alpha-emitting contamination.

6

MEASUREMENT
OF RADIATION

STUDY OBJECTIVES

This chapter will enable the student to:

- List different mechanisms for the detection or measurement of radiation.

- Differentiate between active and passive radiation detection devices.

- Identify the types of radiation detection devices commonly used for NORM work.

- Describe a gas-filled detector.

- Explain the gas amplification curve.

- Describe a scintillation detector.

- Describe a thermoluminescent dosimeter.

6.1 MECHANISMS FOR RADIATION DETECTION

Many mechanisms exist for detecting different types of radiation. Changes can be measured and related to radiation exposure or dose in any of the following ways:

BIOLOGICAL	Changes to a biological system resulting from radiation exposure are measurable. Accuracies are variable, and only very large doses may be effectively quantified.
IONIZATION	The ionization of a gas may be measured either in terms of quantifiable ionization or by counting ionization events.
CHEMICAL	Chemical changes occurring as a result of free radical production may be used to alter the color or other properties of a chemical. The exposure of photographic film by ionizing radiation is an example of this.
HEAT	The energy deposited by radiation will result in an increase in temperature. This effect is usually unmeasurably small yet it is the principle upon which a nuclear reactor operates. However, the radiation levels in the core of a reactor are many, many orders of magnitude higher than those encountered elsewhere.
SCINTILLATION	Radiation may impart some of its energy to certain materials, causing them to emit a flash of light. The flash of light can then be measured.
THERMOLUMINESCENCE	Energy from radiation stored in the electron configuration of some types of crystalline compounds can be released as a flash of light upon heating the compound.

Radiation cannot be detected by human senses. For this reason, we must have something that responds to radiation in some fashion (a detector) and a system to measure the extent of that response (measuring equipment). There are many types of detectors being used to detect radiation. A discussion of the principles of operation of all these devices is beyond the scope of this text.

Radiation detection devices can be either active or passive. All the radiation survey instruments that are used for radiation surveying in the oil industry are active devices, i.e., their output is dependent upon modification, processing, and/or amplification of the direct effect caused by the radiation they measure. Active radiation detection instruments consist of three basic components:

- A detector that will exhibit some measurable change in its properties or generate a signal in the presence of radiation.

- A readout device to display or record the response of the instrument.

- A power supply and signal amplification system.

Passive devices measure radiation without the need for external power. For practical purposes, passive detectors are used almost exclusively for monitoring radiation dose. They "store" the cumulative effects of radiation exposure and thus can be used to measure total exposure over time, which can be equated to dose. Although there are some active devices for monitoring dose, they actually measure exposure and integrate the measurement over time.

Three types of instruments are commonly used to measure radiation in the oil industry: (1) gas-filled detectors, (2) scintillation detectors, and (3) thermoluminescent dosimeters. Gas-filled detectors measure radiation by measuring the amount of ionization produced in a gas. Scintillation detectors measure radiation by the amount of light given off by radiation interacting with a crystalline scintillant. Thermoluminescent dosimeters can be made to give off a flash of light proportional to the dose they have received.

6.2 GAS-FILLED DETECTORS

One of the most common devices used to detect radiation is the gas-filled chamber. As the name suggests, it is a closed chamber filled with some special gas. The chamber is usually wholly or partly constructed from an electrically conductive material and has an electrode inside it. Often one wall of the chamber will consist of a very thin strong membrane or "window" to allow radiation with low penetrating ability, such as alpha and beta particles, to enter. The chamber is filled with a gas or mixture of gases

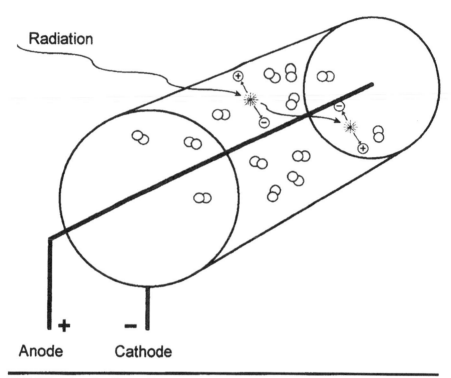

FIGURE 6.1 GAS-FILLED DETECTOR

chosen according to the intended purpose of the detector. The gas(es) may be at atmospheric pressure but are often at a lower pressure and occasionally a higher one. A generic gas-filled detector with a single central electrode is displayed in Figure 6.1.

A voltage is applied to the electrode and the chamber wall such that the electrode becomes positively charged (anode) relative to the chamber wall, which becomes negatively charged (cathode). When radiation passes into the chamber and interacts with the gas, it causes ionization of the gas, which produces positively and negatively charged ion pairs. The electric field between the cathode and anode attracts the ions toward these electrodes. The positive ions are drawn toward the negatively charged wall of the chamber and the negative ions toward the anode. Depending upon the applied voltage, this can cause a very small current of drifting ions to flow between the anode and cathode or it can cause an avalanche of additional ions, resulting in an electrical pulse.

Several factors affect the number of ion pairs and the size of the electrical signal produced:

- The type of radiation passing through the detector.

- The energy of the radiation entering the detector.

- The density of the gas in the detector.

- The type of gas used to fill the chamber.

- The physical size of the detector.

- The voltage applied across the chamber electrodes.

All of the aforementioned factors affect the characteristics and performance of the detector. However, the applied voltage deserves further examination as it fundamentally affects the nature of the detector. Depending upon the voltage, the same detector may be used as an ion chamber, which measures the total deposited energy due to a radiation field, or as a Geiger-Mueller detector, which counts individual ionization events.

6.2.1 THE GAS AMPLIFICATION CURVE

For any gas-filled detector, a relationship exists between the applied voltage across the chamber and the number of ion pairs collected by the detector. This relationship results in a curve or graph called the gas amplification curve (Figure 6.2). This curve is also referred to as the six-region curve because it has six distinct regions associated with it, each of which describes how a gas-filled detector will perform at a particular applied voltage.

It should be noted that the gas amplification curve does not apply to only one detector. Detectors are only designed to operate in one region at one time instead of all six. Each region has its own distinctive characteristics.

RECOMBINATION REGION   Very low applied voltage. Many ion pairs spontaneously recombine before reaching electrodes. The number of ions which do reach the electrodes is very strongly dependent upon the applied voltage. Unusable region.

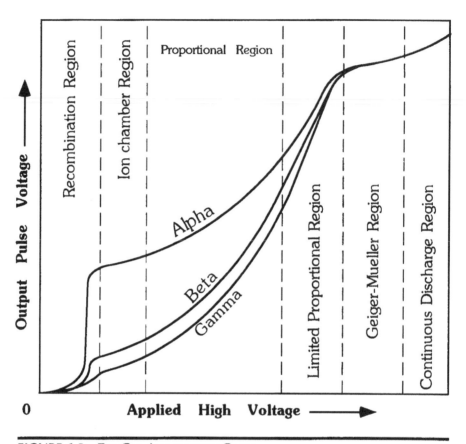

FIGURE 6.2   THE GAS AMPLIFICATION CURVE

ION CHAMBER REGION	Ions attracted toward electrodes. Current generated is proportional to amount of ionization taking place in the gas. Output signal is unaffected by small changes in applied voltage.
PROPORTIONAL REGION	Applied voltage is sufficient to accelerate ions toward electrodes so that they gain sufficient energy to cause secondary ionizations. This effect is termed "gas amplification" and gives an output in the form of electrical pulses, one for each radiation interaction. The pulse amplitude is proportional to the amount of ionization caused

by each interaction and can be calibrated to measure the incident energy of the radiation being measured. This technique is known as pulse height analysis. Because the pulse amplitude will also vary in direct proportion to any change in applied voltage, a steady voltage is required.

LIMITED PROPORTIONAL REGION

Output pulse amplitude increases with incident energy and applied voltage, but not in a constantly proportional fashion. No detectors can usefully operate in this region.

GEIGER-MUELLER REGION

The applied voltage is high enough that any ionization event will cause every single gas atom or molecule to ionize. The output pulse is quite large, but its amplitude is relatively constant with increasing voltage. This means that the detector can only be used to count ionization events, irrespective of the energy or type of radiation interaction causing them. Most simple radiation counters operate in this region.

CONTINUOUS DISCHARGE REGION

The applied voltage is so high that the gas in the tube is ionized without radiation being present. Unusable region.

The gas-filled detector commonly used in the oil field is the Geiger-Mueller (G-M) detector. The design of the probe is common to most manufacturers. It is circular in shape, approximately 2 inches in diameter and 0.5 inches thick. Because of its shape, it is commonly called a "pancake" probe. It is designed so that the measuring face has a surface area of approximately 20 square centimeters. The probe is filled with a blend of gases at below atmospheric pressure. It contains a circular grid cathode. One wall of the chamber consists of an impermeable membrane, sufficiently thin to allow alpha and beta particles to pass through it. This makes the pancake probe the detector of choice for detecting the presence of NORM contamination. When installed in a probe, the membrane is usually protected by a wire mesh. There are some other probes that can perform the function, but they are generally more expensive and fragile.

6.3 Scintillation Detectors

Scintillation is the mechanism by which a material emits light upon interaction with ionizing radiation. For a particular type of radiation, the intensity of the light emitted is related to the energy of the incident radiation (it should be noted that this relationship is not linear). The light can be measured and equated to radiation exposure.

Different types of radiation require different types of scintillating material to detect them. For solid-state detectors, a material called anthracene is commonly used to measure beta radiation. Zinc sulfide is used for measuring alpha radiation, and sodium iodide is used for gamma radiation. In the biological sciences, most analyses involve weak alpha- and beta-emitting samples of radioactive material. These are mixed with a scintillating material in a liquid solvent. This technique is known as liquid scintillation.

The type of solid-state scintillation detector most commonly used for NORM surveying is illustrated in Figure 6.3. This type of scintillation detector contains a 1-inch-long, 1-inch-diameter cylindrical crystal of sodium iodide, which has a tiny amount of thallium in it to improve its scintillating properties. The crystal is enclosed by aluminum, which reflects light back into the crystal and protects it from moisture that would rapidly destroy it.

Ionizing radiation interacting with a scintillation crystal excites its electron structure. The resulting excess energy is released by the emission of photons. These photons constitute a very faint flash of visible light, which must be accurately measured. This is accomplished by the use of a photomultiplier tube (PMT). The PMT converts the minute light pulse into an electrical pulse.

The flash of light from the scintillation crystal is admitted into the PMT by a quartz window in the end of the crystal housing. The light strikes a thin layer of a photosensitive substance such as cesium antimonide, known as a photocathode. It emits electrons when struck by a photon. Close to the photocathode is the first of a series of collecting electrodes called dynodes. The detector may have as many as nine or ten dynodes.

Electrons from the photocathode are attracted by the first dynode, which has a positive charge. By the time the electrons get to the first dynode, they have acquired enough energy to free several more electrons from the dynode surface. These electrons are attracted toward the next dynode, which has a greater positive charge. When all these electrons reach the next dynode, they, in turn, will have acquired enough energy to dislodge more electrons. In this way, the number of electrons increases or multiplies. From

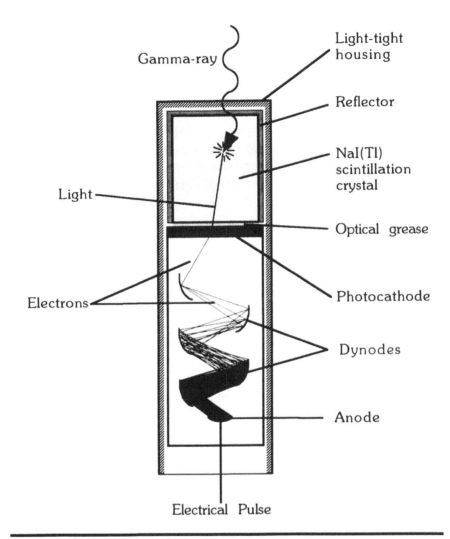

FIGURE 6.3 SCINTILLATION DETECTOR

the last dynode, the electrons are attracted to and collected at the anode of the tube. The current that is measured is converted to a meter reading. The energy of the original gamma ray is typically multiplied one million times by the time it reaches the anode.

The crystal and photomultiplier tube are encased together, usually in aluminum. This is done to provide shock resistance for these fragile items and to prevent outside light from entering the photomultiplier tube.

6.4 Personnel Monitoring Devices

External radiation fields are measured mainly by the use of electronic devices. These devices measure the field of exposure to which our bodies are being subjected, but they do not measure what we are most concerned with—radiation dose (i.e., how much of this radiation is actually interacting with our bodies and causing biological damage). The process of measuring radiation dose is called dosimetry. Dose is measured using devices called dosimeters. Only two types of dosimeters are in regular use in the oil industry: the film badge and the thermoluminescent dosimeter (TLD). Of these, the TLD is the overwhelming favorite.

6.4.1 Film Badges

The film badge can provide a reasonably accurate, permanent record of the cumulative exposure received by an individual while working in a field of radiation. The most commonly used badges measure both beta and gamma radiation; many measure neutron doses also.

A typical film badge is composed of a dental-size sheet of photographic film, wrapped in a light-proof package and housed in a plastic cover. When radiation interacts with the film, the film is exposed, in a similar fashion to taking a photograph. When developed, the film will be darkened where ionization has taken place. The degree of darkening is proportional to the radiation dose received by the film.

6.4.2 Thermoluminescent Dosimeters

In the early 1900s, it was discovered that some materials would emit light if exposed to radiation and then heated. This phenomenon was termed thermoluminescence. The light given off during heating is proportional to the amount of radiation to which the material has been exposed. A photomultiplier tube can be used to convert the light into a measurable electrical signal.

TLD material is somewhat similar to scintillating material, except that TLD material will "store" the energy from radiation until heated whereas scintillators give up that energy immediately. The most widely used thermoluminescent materials are lithium fluoride, calcium fluoride, and calcium sulfate. TLD material may be used in powder form but is most widely used in the form of small crystals known as chips. It can also be impregnated into

TABLE 6.1 ADVANTAGES AND DISADVANTAGES OF FILM BADGES VERSUS TLDs

TLD	Film Badge
Reusable—upon heating, it is "reset" and may be used again	Each piece of film is used only once
No hard copy record	Developed film is a permanent record of dose received
Once read, the information is lost (i.e., a TLD reading must be obtained correctly the first time)	Provided film is developed correctly, it may be "read" to obtain a dose reading many times
Rugged and stable	Somewhat susceptible to heat and moisture
Sensitivity about the same as film (some new types may be ten times as sensitive in a convenient form)	Sensitivity about the same as TLDs (more sensitive film is too unstable)

silicone-based material or PTFE (Teflon). A TLD badge is the most common means of dosimetry among oil field workers. In some industries where radioactive materials are literally handled with the hands, it is common to find TLDs incorporated into a finger ring.

Some factors to consider in the use of TLDs versus film badges are listed in Table 6.1.

7

SURVEYING AND SAMPLING NORM

STUDY OBJECTIVES

This chapter will enable the student to:

- List the reasons for performing radiological surveys.
- Describe the pre-operational checks that must be performed on survey instruments.
- List some considerations when performing radiation surveys.
- List some considerations when performing contamination surveys.
- Define counts per minute (cpm) and decays per minute (dpm) and convert between them.
- Understand the process of whole-body frisking.
- Describe the basic process of personnel decontamination.
- Understand the means of screening for NORM-contaminated soil and land.
- Discuss the technique used to determine if items of equipment or tubulars are NORM contaminated.
- Name the units used to measure airborne radioactive contamination.
- Name the units used to regulate airborne radioactive contamination.
- Identify when an Airborne Radioactivity Area should be posted.

7.1 INTRODUCTION

Surveys are performed for a number of reasons. Foremost, they tell us the radiological conditions of the area in which work is taking place. They also tell us where and whether to post warning signs. In addition, surveys are performed to:

- Determine proper radiological controls.

- Keep workers informed of radiological conditions.

- Determine whether or not dosimetry is required.

- Identify licensing requirements.

- Identify the need for respiratory protection.

- Determine the effectiveness of decontamination.

- Determine if previously contaminated equipment and areas can be released for unrestricted use.

Surveys may consist of measuring exposure, soil contamination, and airborne and waterborne radioactivity. Just as important as obtaining survey data is the interpretation of that data. A radiation surveyor must also know the history of the equipment or area being surveyed. This gives the surveyor the ability to decide whether or not a particular piece of equipment may be contaminated.

7.2 CHECKING AND TESTING SURVEY INSTRUMENTS

Prior to using any portable survey instrument, pre-operational checks should be performed. For NORM survey use, instruments are generally required to be calibrated every 12 months (although some states may have other requirements). It is possible that the condition of the instrument could change between calibrations due to damage, improper maintenance, or other reasons.

Four basic checks should be carried out every day that a survey instrument is used or following any incident that could possibly cause damage. If a surveyor is traveling any distance to perform a survey, it is wise to check the instrument before leaving and again following the journey. The four checks are as follows.

7.2.1 Calibration Check

Before even switching the instrument on, the calibration sticker should be checked to ensure that the instrument is within its valid calibration dates. A sticker bearing the calibration date and recalibration due date is usually affixed to the side of the instrument. If the sticker is missing, the instrument should be set aside until the calibration file can be checked or the instrument has been recalibrated.

7.2.2 Physical Inspection

Still without turning the instrument on, it should be briefly inspected for any physical damage or defect. Checks should include:

- Check the meter face. Make certain the glass is secure and not cracked. The needle should be straight and must not touch the glass or meter face.

- Inspect the meter housing and ensure it is in good shape. Make sure the handle is firmly attached. Shake it *gently* to make sure there is no water or loose parts rattling around inside.

- Inspect the cable for cuts, abrasion, or severe kinks. Ensure that the connectors are not loose and have no moisture inside them.

- Ensure that the probe is not damaged or broken. Shake it; there should be no sound (i.e., no bits rattling around inside). For a pancake probe, blow gently onto the face of the probe; there should be no metallic rattle. Make sure there is no mud or grit on the fragile surface. If so, do not attempt to clean it; replace it with a clean one.

7.2.3 Battery Check

The condition of the battery should be determined each time the instrument is used. Battery life is long enough to preclude the necessity of frequent battery checks while using most instruments. However, when performing lengthy surveys, it is a good idea to perform a battery check periodically. When performing a battery check, turn the main rotary switch to the "BAT" position for at least 10 seconds (old batteries that have not been used for some time will appear strong but will weaken rapidly). The needle should hold steady in the "BAT TEST" region of the display. The battery check display for a typical survey instrument is shown in Figure 7.1.

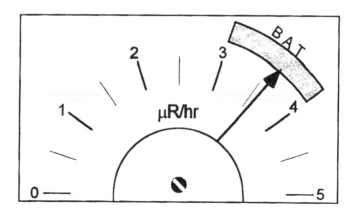

FIGURE 7.1   BATTERY TEST O.K.

7.2.4 RESPONSE TEST

The response of the instrument should be checked against a known source. The source designated by company procedures should be obtained and the reading checked against that particular source. The lowest scale multiplier setting possible, without the reading being off-scale, should be used. This check should be performed in an area of low background.

The reading should be within ±20% of the normal figure for that particular instrument and source. Any instrument that fails this test should be immediately removed from service until repaired and/or recalibrated.

Whenever a new instrument is received or an instrument is returned from calibration, its reading to a known source will have been determined. This is the "normal" reading mentioned above.

7.2.5 POST-SURVEY CHECKS

After a survey instrument has been used, the battery test and response test must be carried out again. Successful completion of these tests will ensure that any survey data recorded between the pre- and post-survey checks are valid (probably).

Whenever instrument readings will constitute a part of any documentary record, a note confirming the performance of pre- and post-survey instrument checks and their outcome must be included with the document. (Remember to include the calibration due date and instrument serial number.)

7.3 Maintenance and Care of Survey Instruments

Every manufacturer of field-portable radiation survey instruments makes models which have suitable capabilities for NORM surveying. Those which are most commonly used are all fairly robust, but they are by no means indestructible and certainly not waterproof (or idiot-proof for that matter). When reasonably well cared for, these instruments have proven to be extremely reliable. With appropriate precautions, they can be used in all weather conditions.

7.3.1 Pre-Survey Checks

Carrying out the pre-survey checks described in Section 7.2 will ensure that any problems are rapidly discovered. This will minimize the time needed to get the instrument back in good working order.

7.3.2 Probe Compatibility and Interchangeability

Most instruments equipped with scintillation probes have a calibrated readout; that is, they give a direct reading in microroentgens per hour (μR/hr) as opposed to counts per minute (cpm) of a "frisker"-equipped instrument. Each scintillation detector is slightly different, and two different probes cannot be expected to give the same reading when exposed to the same source. Therefore, it is *never* acceptable to switch scintillation probes from one instrument to another. If you have two such instruments, one with a "bad" probe and the other with a "bad" main instrument section, then you have two bad instruments.

The Geiger-Mueller (G-M) pancake probes are all manufactured to be fairly similar in response for a given operating voltage. Because the operating voltage is determined by the main instrument section, any probe should give a similar reading with the same instrument utilizing a counts per minute readout. If it becomes necessary to switch probes, a response test should be carried out to ensure that the probes are indeed compatible.

It is never acceptable to use a G-M-type probe with an instrument that is set up for a scintillation detector and vice versa. While the instrument will "work" (i.e., it will make a noise and the needle will move in response to radiation), any measurements made will be worthless.

Cables can be freely switched between instruments (the instrument should be switched off before disconnecting a cable to avoid instrument damage or electric shock). The main limitation is cable length, which cannot normally exceed 40 feet, although this should not be a problem in most cases. Connector compatibility could also be a problem, as several different designs are in use. Only quality co-axial cables should be used.

7.3.3 Radiation Survey Instruments and Water

The commonly used field-portable survey instruments are not waterproof. Some manufacturers produce so-called "environmentally sealed" versions with rubber covers over the switches. These are generally only marginally effective in keeping moisture out, although they may help to keep dust out (and moisture in). Most instrument manufacturers now produce sophisticated models with digital readouts and other advanced capabilities. Such instruments are more easily damaged by moisture and considerably more expensive to repair.

The operating voltage of the probes is typically between 600 and 1000 volts. If moisture gets into the probes, cables, cable connectors, or into the main instrument housing, it will short out this high voltage very easily (it may also give the operator a nasty shock).

Should an accident occur and water get into the instrument, whether or not its operation appears impaired, it should immediately be switched off and the batteries removed. It should be given to the radiation work supervisor or other person responsible for instrument repair. As long as a wetted instrument is not switched on, it can often be salvaged fairly easily. Salt water will do much more damage than fresh; therefore, if an instrument becomes soaked with salt water, it often helps to disassemble it as soon as practicable and wash it in fresh water or alcohol. This should not, however, be done without first consulting the radiation work supervisor or other person responsible for instrument repair.

If a survey must be carried out in the rain or under other adverse conditions, judicious use of sturdy plastic bags, such as one-gallon resealable freezer bags (great for taking samples too!) and electrical tape can waterproof a survey instrument fairly well. This will not affect the operation of a scintillation detector, but a pancake probe cannot be covered and still detect alpha and beta radiation. Special waterproofed scintillation probe and cable assemblies, which can actually be used underwater, are available. They can be rented for special-purpose surveys. When using a "frisker" in wet weather, there is no substitute for sheltering it from the rain.

7.3.4 Do's and Don'ts of Instrument Care

- *Do* ensure that instrument calibrations are kept up to date.

- *Do* use only high-quality alkaline batteries (rechargeable batteries can sometimes be used—check with the manufacturer). Ensure that they are inserted properly. Remove them if the instrument will not be used for any length of time.

- *Do* perform instrument checks before and after use for surveying, transportation, rough treatment, or any other suspected damage.

- *Do* remember to obtain the correct check source before taking an instrument out in the field.

- *Do* cover the instrument and probe (scintillation probes only) with plastic bags in wet weather.

- *Do* be careful with fragile electronic equipment. Scintillation probes are especially susceptible to sharp blows.

- *Do* switch probes between "friskers" if required, but perform a response check to ensure accuracy before using.

- *Do* switch out cables if required.

- *Do not* switch probes between microroentgen survey meters or from microroentgen to "frisker"-type instruments or vice versa.

- *Do not* loosen or detach cable connectors with the instrument switched on.

- *Do not* allow the meter needle to go off-scale (peg-out) for any longer than absolutely unavoidable.

- *Do not* use an unprotected instrument in the rain or near water spray, especially salt water.

- *Do not* continue to use an instrument if it has become wet.

- *Do not* bother performing a survey with a meter that is out of calibration or has not been checked. The survey will probably have to be done over again.

7.4 PERFORMING RADIATION SURVEYS

Some of the more common categories of surveys are confirmatory surveys, pre-job surveys, job coverage surveys, and post-job (or release) surveys. Surveys may also be performed as the result of a casualty (i.e., spill) or when radiological conditions are expected to have changed. Some of the techniques that should be utilized when performing radiation surveys are as follows:

- Prior to performing a survey, background radiation in the area should be measured. Most states that regulate NORM now use absolute limits rather than subtracting background. However, it is useful to establish the level above which NORM should be considered present. The background reading should be conducted at or beyond the site boundary. This ensures that the presence of any NORM does not influence the background reading. Many factors can influence background radiation levels, such as faulting, ground cover, certain high-potash fertilizers, shale cuttings, etc. Therefore, if in any doubt, extend the area of measurement.

- Protective clothing and dosimetry should always be considered when entering an area of unknown radiological conditions. Shoe covers, or rubber boots and gloves, as a minimum, should be worn. Protective clothing should be consistent with that worn by other workers in the area.

- When performing radiation surveys on equipment and tubulars, it is important to ensure that the probe is held within 1 centimeter (0.5 inches) of the surface. Because the dose rates associated with NORM-contaminated equipment are so low, surveying at distances greater than 1 centimeter can result in substantially lower readings than those actually present.

- Most survey instruments are equipped with an audible response. The audible response should always be used when surveying for NORM. This is because the audible response to changes in radiation exposure is immediate, whereas the meter response is much slower. When radiation levels increase, the audible response will give a more rapid indication than the meter needle movement.

- Equipment and tubulars that are stored in close proximity to each other should be separated. This prevents the radiation from one piece of equipment from interfering with the reading on another. Examples include surveying a stack of tubulars or a group of drums stored together.

- On many common survey meters, 2-meter response settings are available. The fast response setting allows 90% full-scale reading in approximately 4 seconds. On slow response, 22 seconds is required to reach the same 90% reading. The fast mode should always be used when performing surveys for NORM. The survey rate should be slowed when an increase in audible response is heard. This allows a more accurate determination of dose rates to be made.

- General area, background, and boundary surveys should be performed with the probe held at waist height. Audible indication should be on and fast response should be used for these surveys.

- When surveying tubulars, the insides of the pipe should always be checked by placing the probe a few inches inside each end. This is particularly important when releasing them for use in an unrestricted area. This is not required when plugs or end caps are in place or when radiation levels clearly indicate the presence of NORM.

- When performing release surveys on equipment, it is important to ensure that 100% of all accessible surfaces are surveyed.

- When surveying for NORM, areas where NORM is most likely to be found should always be surveyed first. Such areas include tubulars, tanks, heater treaters, and separators. Any area that might cause NORM to collect (e.g., drain valves, bends in pipe, tank bottoms) should be carefully surveyed.

- Always draw a map. Do not rely on memory to provide the recall necessary for a complete and thorough survey. A rough sketch is all that is necessary while performing the survey. If desired, a more professional map can be completed after exiting the restricted area or leaving the site.

- When screening soil and land for NORM, some state guidelines require that the probe be held within 2.5 centimeters (1 inch) of the surface. However, when surveying large areas, it is probably wiser to hold the probe at mid-shin height (3 to 9 inches) so as to reduce the chance of breakage by impact from stones, pipes, etc.

- Soil and land should be marked as potential sampling locations if readings are twice background or greater. This is strictly a screening procedure and not a determination that the soil is contaminated. Whether or not soil contamination is above regulatory limits can only be determined by laboratory analysis of a soil sample.

An example of a survey form intended for documenting simple NORM surveys is provided in Figure 7.2. Any similar format can be used provided it contains all the relevant information.

7.5 PERFORMING CONTAMINATION SURVEYS

It is important to note that some state regulations require contamination surveys to be performed and may stipulate the techniques that should be used.

Surveys for loose surface contamination are performed using 1-inch-diameter cloth disks called smears. The smear should be wiped over an area of approximately 100 square centimeters. This corresponds to a 4-inch-by-4-inch square or a line 16 inches long. The activity of the smear can then be determined using an instrument outfitted with a pancake probe. If the smear is held within 0.5 inches of the probe for 5 seconds and no sustained counts above background are detected, the surface from which the smear was taken can be considered to be free of loose contamination. If increased counts are detected during the first 5 seconds, the smear should continue to be counted until the meter stabilizes. It may be necessary to use the slow response setting on the instrument to obtain an accurate reading.

When surveying large areas or pieces of equipment, the large area smear (LAS) technique may be used. A LAS is a general wipe-down of a relatively large area. It can be accomplished using a rag or sturdy paper towel. Large parts of potentially contaminated surfaces can be wiped down and then checked for contamination using a pancake probe. The LAS technique is ideal for releasing items that are probably not contaminated, such as cleaned-out drums, hand tools, or large areas of deck. It should be noted, however, that although this technique is very sensitive, it may be too informal to meet many regulatory requirements. Also, this technique is too unscientific for any results other than "no detectable contamination" (N/D) to be used for most legally recognized release surveys. If contamination is detected, the item should be decontaminated and remeasured or a further contamination survey should be performed using cloth patches to determine whether the level of contamination is of regulatory concern.

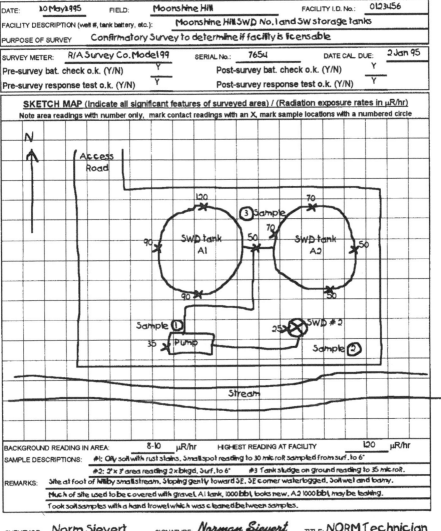

 **XYZ PETROLEUM**

NORM SURVEY RECORD

DATE: 10 May 1995   FIELD: Moonshine Hill   FACILITY I.D. No.: 0123456

FACILITY DESCRIPTION (well #, tank battery, etc.): Moonshine Hill SWD No. 1 and SW storage tanks

PURPOSE OF SURVEY   Confirmatory Survey to determine if facility is licensable

SURVEY METER: R/A Survey Co. Model 99   SERIAL No.: 7654   DATE CAL. DUE: 2 Jan 95

Pre-survey bat. check o.k. (Y/N)   Y   Post-survey bat. check o.k. (Y/N)   Y

Pre-survey response test o.k. (Y/N)   Y   Post-survey response test o.k. (Y/N)   Y

SKETCH MAP (Indicate all significant features of surveyed area) / (Radiation exposure rates in μR/hr)

Note area readings with number only, mark contact readings with an X, mark sample locations with a numbered circle

BACKGROUND READING IN AREA: 8-10   μR/hr   HIGHEST READING AT FACILITY   120   μR/hr

SAMPLE DESCRIPTIONS: #1: Oily soil with rust stains. Small spot reading to 30 micro R. Sampled from surf. to 6"

#2: 2' x 3' area reading 2 x bkgd. Surf. to 6"   #3 Tank sludge on ground reading to 35 micro R.

REMARKS: Site at foot of hill by small stream. Sloping gently toward SE. SE corner waterlogged. Soil wet and loamy.

Much of site used to be covered with gravel. A1 tank, 1000 bbl, looks new. A2 1000 bbl may be leaking.

Took soil samples with a hand trowel which was cleaned between samples.

SURVEYOR: Norm Sievert   SIGNATURE: *Norman Sievert*   TITLE: NORM Technician

REVIEWED BY: *Ray Nium*   SIGNATURE: *Raymond Nium*   TITLE: RSCO

FIGURE 7.2   SAMPLE NORM SURVEY SHEET

In some circumstances, if loose contamination is indicated on a smear, it may be necessary to quantify the alpha- and beta-emitting components of that contamination. To accomplish this, a single sheet of thin paper may be held between the smear and the probe. The paper will shield all of the alpha particles so that the reduction in count rate may be considered to be the count rate due to alpha radiation.

The activity of a smear should be reported in units of decays per minute (dpm) or counts per minute (cpm) over the area the smear was taken (dpm/ 100 square centimeters or cpm/LAS). If the area of the smear is less than 100 square centimeters, it should be reported as dpm/smear.

7.6 COUNTS PER MINUTE VERSUS DECAYS PER MINUTE

The amount of radioactivity contained in a particular substance can be quantified using several terms. Three commonly used terms are applicable to the oil industry:

1. The curie (Ci) or becquerel (Bq)

2. Decays per minute (dpm)

3. Counts per minute (cpm)

The curie and becquerel are units of measurement used to quantify the radioactivity of a material. They are often expressed in units of decays (disintegrations) per second (dps) or disintegrations per minute (dpm).

A curie is defined as the amount of radioactive material that decays at the rate of 3.7×10^{10} dps or 2.22×10^{12} dpm. A becquerel is defined as the amount of radioactive material that decays at the rate of 1 dps. This is, of course, equal to 60 dpm.

Disintegrations per minute is a measure of the actual nuclear events or decays taking place each minute inside a source material. Counts per minute is a measure of the actual number of decays being detected by a given radiation detection device.

There is an important and fundamental difference between counts per minute and disintegrations per minute. A particular radiation detection instrument will only "see," and therefore be able to count, a fraction of the actual radioactive decays occurring in any radioactive substance. A proper explanation of this phenomenon depends on many factors beyond the scope of this text. However, the proportion counted (cpm) can be related to the decays actually taking place (dpm) through the instrument's efficiency factor.

The disintegrations per minute activity of a source can be determined by dividing the counts per minute by the efficiency of that particular instrument. Thus:

$$dpm = \frac{cpm}{efficiency}$$

Most contamination survey instruments read out in counts per minute. In order to calculate the disintegrations per minute activity in a particular source, the efficiency of the detector must be known. To further complicate matters, the efficiencies of any given type of probe are different for different types of radiation. The efficiencies for the pancake probe are as follows:

Alpha	30% (0.3)
Beta	10% (0.1)
Gamma	0.01% (0.0001)

The figures quoted are generalized approximations only. In fact, the efficiency varies according to the energy of the particles and many other factors. In the field, the information required to derive the exact efficiencies is not available, and the approximations are, therefore, always used.

From the above figures, it is apparent that the G-M pancake probe is so inefficient at detecting gamma rays that, for all intents and purposes, it can be considered an alpha and beta detector only. Note that this is definitely not true for all G-M type probes.

Efficiency factors allow the conversion of counts per minute to disintegrations per minute, e.g., if a reading of 100 cpm was obtained on a particular source and it was determined that it was all due to beta radiation, the disintegrations per minute taking place in the source would be determined as follows:

$$dpm = \frac{cpm}{efficiency} = \frac{100}{10\%} = \frac{100}{0.1} = 1000$$

Similarly, if a reading of 60 cpm was determined to be due entirely to alpha radiation, the disintegrations per minute in the source could be calculated as follows:

$$\text{dpm} \quad \frac{\text{cpm}}{\text{efficiency}} = \frac{60}{30\%} = \frac{60}{0.3} = 1000$$

When performing measurements in the field, it is permissible to round off calculations. For instance, 25/30% can be exactly expressed as 25 × 3.333, but it is much easier to round off and calculate alpha disintegrations per minute by multiplying the counts per minute figure by three.

7.7 MONITORING PERSONNEL FOR CONTAMINATION

As Low As Reasonably Achievable (ALARA) policy requires all personnel to be personally responsible for their own radiation exposure. This holds true for contamination control as well. Anyone who has been working in a restricted area should, as a minimum, perform a hand and foot frisk prior to exiting the area. If the person has been working in an area where loose surface contamination is known to exist or if that person has been wearing protective clothing, a whole-body frisk should be performed. This should be accomplished using the G-M pancake probe. The pancake probe is often referred to as a "frisker" for this reason.

Background count rate in the area where frisking is performed must be considered. The background rate should not be greater than 50 cpm. When performing personnel monitoring, the audible response should always be used. Any increase in audible count rate indicates the presence of contamination. The slow response mode should also be used when performing personnel frisking. On instruments outfitted with an alarm, the alarm should be set to trigger at 100 cpm.

The hands should be frisked first before touching anything else. This prevents contamination of the probe should the individual's hands be contaminated. Body contact with the probe should be avoided. Once the hands have been frisked, the probe may be picked up. The frisk is performed by holding the probe within 0.5 inches of the surface and moving it at a rate of 2 to 3 inches per second. Certain areas of the body require special attention, including:

- The head and hair.

- The face, especially around the mouth and nostrils, paying special attention to any facial hair.

- Hands and feet, including cuffs and ankles of clothing.

- Knees, elbows, and forearms.

- The buttocks.

An appropriately thorough whole-body frisk should take at least 3 minutes. If problems or questions arise, the radiation work supervisor should be contacted.

It is important that the audible response be used when conducting personnel frisking. Any increase in the audible rate indicates the possible presence of contamination. If an increase is detected, the frisking rate should be slowed and the area of the highest count rate determined. Any sustained counts above the background reading indicate the presence of contamination. The radiation work supervisor should be contacted for advice on the proper decontamination method. Only those personnel with a genuine medical emergency will be allowed to exit a restricted area while contaminated.

7.8 Personnel Decontamination

When the decontamination of personnel is required, the method used should be chosen on the basis of not only the effectiveness of removing the contamination, but also the effect it will have on the individual.

NORM contamination of the skin is best removed by washing the affected area in tepid water. A mild, high-lather soap may be used, especially if the NORM material of concern is in an oily or greasy form. Scrubbing with a soft bristle brush should be performed only with the express permission of the radiation work supervisor.

If hot water were to be used for removing NORM contamination from the skin, the pores would open up and could let the contamination penetrate deeper into the skin. Conversely, if cold water were to be used, it would cause the pores to close, possibly trapping contamination in the skin.

Once the washing process is complete, the area should be surveyed again to ensure the contamination was completely removed. The washing process should be performed a maximum of three times. If contamination is still present after three washings, then washing with soap may be tried (if not already done). If this is also unsuccessful, then the radiation work supervisor should be contacted for further guidance.

Contaminated clothing, gloves, and rags should be placed in a 55-gallon DOT shipping drum or similar suitable container for disposal. If required, clothing, cotton gloves, and rags may be washed in a washing machine for re-use or disposal as "clean." Of course, this will require a dedicated washing machine and the careful handling of wastewater.

Any water used to decontaminate personnel or clean clothing should be collected and held for laboratory analysis before releasing it to an unrestricted area. If sufficient contamination existed on the individual or clothing, the NORM content of the water may very well exceed regulatory limits.

7.9 SAMPLING NORM

When screening soil and land suspected of being contaminated with NORM, areas indicating radiation levels greater than twice background should be marked as potential sampling locations. It should be noted that this is only a screening procedure and not a method of determining whether or not soil is actually contaminated. That can only be determined by laboratory analysis of the samples.

When sampling soil or other ground-covering materials, the samples should be taken to a depth of 15 centimeters (6 inches). This may be accomplished with a pickax, shovel, power auger, or other suitable equipment. A minimum of about 1 kilogram (2 pounds) of material should be collected. Gallon or half-gallon resealable bags provide an excellent sample container (if soil is wet or otherwise unpleasant, samples can be double-bagged for extra security). The bag should be filled approximately three-quarters full and marked with the appropriate information. The sample container should be properly sealed and labeled with the date, time, sample location, and some form of unique identification. The soil must be sent to a qualified laboratory for analysis. When taking soil samples, rubber gloves should be worn to minimize the potential for hand contamination.

Liquid and sludge samples should be placed in durable, non-breakable containers capable of holding approximately 4 liters (1 gallon). Thoroughly cleaned plastic milk containers work well. Otherwise, wide-mouthed, screw-top plastic jars can be obtained expressly for this purpose. The container should be sealed and marked with the sample date, time, and location. A brief description of the contents of the container should also accompany the sample to the laboratory.

Additional considerations for sampling include:

- Never send liquid or sludge samples to the laboratory in glass or other breakable containers.

- Always wipe down the outside of sample containers and perform smears to ensure the outside is not contaminated. Department of Transport (DOT) regulations require this if samples are to be shipped.

- Take representative samples of the whole area. Take as many samples as necessary, commingling samples when appropriate.

- When sampling drums or containerized NORM, be sure to sample the part of the container that shows evidence of the highest contamination.

- Determine background levels by sampling the soil outside the area where work will be performed.

7.10 SAMPLING FOR AIRBORNE RADIOACTIVITY

Air sampling is performed to determine the concentration of radioactive material present in the air. This information is required for the following reasons:

- To enable a determination of the effectiveness of engineering controls and practices designed to reduce airborne contamination.

- To determine the amount of airborne radioactive material to which personnel working in a restricted area are being exposed.

- To assist the radiation safety supervisor in deciding whether personnel should be using respiratory protection apparatus and, if so, to make an appropriate selection from the devices available.

- To measure the amount of radioactive material being blown out of a restricted area by the wind.

The amount of radioactive material in the air is measured in units of microcuries per milliliter (μCi/ml) or becquerels per milliliter (Bq/ml). If the amount of radioactive material in the air exceeds regulatory limits, then the area must be posted as an Airborne Radioactivity Area.

Sampling for airborne radioactivity is performed whenever work that may cause radioactive material to become airborne is being conducted. Air sampling should be conducted in the following locations:

- At the boundary of a restricted area in the predominant downwind direction—This may, of course, change during the duration of a job.

Several locations should probably be selected and samples taken at each.

- In the worker's "breathing zone"—This is the area in the immediate vicinity of the worker's face from which material in the air is likely to be inhaled.

Sampling may be performed at other locations if the radiation safety supervisor deems it necessary.

Air sampling is accomplished using a calibrated pump system to draw air through a sample filter. To obtain valid results, a minimum volume of 1000 liters (35 cubic feet) of air must be drawn through the filter. This is typically done by running the sampler at 30 liters per minute (lpm) for 30 minutes. Longer run times and larger volumes may be collected at the discretion of the radiation work supervisor.

After the air sample has been collected, the filter holder should be removed or exchanged for a fresh one, and the filter should be removed with clean tweezers. The use of tweezers prevents the sample from becoming cross-contaminated (only a tiny amount of contamination from the fingers can drastically affect analysis results). The filter should then be placed in a plastic bag and taken away for analysis. The outside of the bag should be marked with the date the sample was taken, the start and stop times, the sample flow rate (cfm or lpm), the location from which the sample was taken, and the printed initials of the person taking the sample.

8

PERSONNEL PROTECTION FROM RADIATION AND CONTAMINATION

STUDY OBJECTIVES

This chapter will enable the student to:

- Define ALARA and appreciate its implications.

- List and explain the components of the tripartite principle used to minimize external exposure to radiation.

- List three basic rules for protection against contamination.

- Discuss the purpose of protective clothing.

- List and discuss some considerations for selecting protective clothing.

- List and discuss some considerations for using protective clothing.

- Describe the process of removing protective clothing.

97

- Discuss the reduction of airborne radioactivity by engineering controls.

- Name the terms used when quantifying airborne radioactive material and its uptake.

- List some of the considerations in the use of respiratory protection equipment.

- Understand the two categories of emergencies involving radioactive materials.

- State the basic principle applied to accident response when working with NORM.

- State and explain the acronym used to remind us of the action for spill response.

- List some considerations in responding to a fire involving NORM.

- List some considerations for responding to an injury accident involving NORM.

- Describe the assessment of external and internal committed doses.

- State the maximum allowable whole-body radiation dose (total effective dose equivalent).

- State the maximum allowable dose to the fetus of a declared pregnant woman.

- State and discuss some of the benefits of training in the principles and practices of radiation protection from NORM.

8.1 INTRODUCTION: THE ALARA PRINCIPLE

In the BEIR III report, the Committee on the Biological Effects of Ionizing Radiation (BEIR) cited evidence that an increased risk of cancer is the principal potential somatic effect of chronic low-level radiation exposure. The report states that there is an estimated increase of between 3 and 8% in cancer mortality resulting from a continuous lifetime radiation exposure of 1 rem per year. Considering this information, the health physics community recognized the need to reduce radiation exposures and thus reduce the possibility of radiation-induced biological effects. The present approach is to reduce exposure to "as low as reasonably achievable" (ALARA). All basic concepts and methods to reduce radiation exposures are based upon this philosophy.

For many years, the federal government has recommended that the ALARA philosophy be employed. It is now mandatory. Radiation protection regulations now specifically stipulate that: "Each licensee shall ensure that the dose to individuals receiving occupational doses and to members of the public is **as low as reasonably achievable...**"

The principle of ALARA is fundamental to all aspects of radiation protection programs designed to reduce the health risk to workers, the public, and the environment arising from exposure to sources of ionizing radiation. ALARA is therefore the major factor in designing programs to protect workers, the public, and the environment from exposure to NORM.

8.2 Protection from External Radiation Exposure: Time, Distance, and Shielding

As applied to protection from external penetrating radiation, the tripartite principle of time, distance, and shielding is quite simple—but it works! The use of these three concepts is as follows:

TIME

Minimize the time spent close to a source of ionizing radiation. This is probably the most basic of all radiation-exposure control techniques. The total dose a person will receive while in the vicinity of a source of ionizing radiation is directly proportional to the length of time for which he or she is exposed. Therefore, ALARA requires that no one spend any more time than absolutely necessary near sources of radiation. ALARA requires the minimization of the dose to each worker. Thus, it becomes desirable to cycle workers through the tasks from which they would receive the greatest radiation exposure. The theory is that it is better for several workers to receive a small dose than for one worker to receive a large dose.

DISTANCE

Maximize the distance between personnel and any sources of ionizing radiation. Radiation exposure decreases rapidly as one moves farther from the source. In the case of a point source (small in relation to the distance from it), the exposure will decrease in inverse proportion to the distance from the source (e.g., the radiation exposure at 6 feet from a point source is only one-quarter of that at 3 feet.) Applying ALARA, one should remain as far away as practicable from any significant source of ionizing radiation.

SHIELDING   **Maximize the shielding between personnel and sources of penetrating radiation.** If shielding is placed between a source of radiation and an individual, the shield will absorb or reflect some of the radiation. The amount of radiation that will pass through depends upon the penetrating ability of the radiation and the properties of the shielding material. This concept has not been deliberately employed to protect personnel from radiation from NORM. However, the steel from which most oil field equipment is constructed provides adequate shielding protection from the relatively low exposures due to NORM. Drums of pipe scale are likely to generate a more significant exposure hazard than any other commonly encountered form of NORM. When many drums are in storage, those that exhibit the highest exposure rates can be surrounded by drums that show lower readings. This utilizes the ALARA concept of shielding since the lower dose rate drums effectively shield personnel from the higher dose rate drums in the middle.

These three principles are illustrated in Figure 8.1.

Except in the case of shielding containers of concentrated NORM material, as described above, or under very unusual circumstances, time, distance, and shielding need not be of too much concern to most personnel working with NORM. However, it is not uncommon for oil field workers to encounter well logging or radiography operations in progress. The sources utilized in these operations are sealed sources which generally pose no contamination problem but emit large amounts of gamma or neutron radiation. Typically, such sources may have an activity from 0.5 curies up to 200 curies (18.5 gigabecquerels to 7.4 terabecquerels) and have the potential to deliver massive doses of radiation in very short periods of time. The personnel trained to handle such sources go to seemingly elaborate lengths in their use of remote handling devices, shielded "pigs," and carefully planned, speedy (but not hasty) operations. They derive all of their radiation protection procedures from the tripartite principle of time, distance, and shielding.

TIME: Minimize the time spent in proximity to a source of penetrating radiation

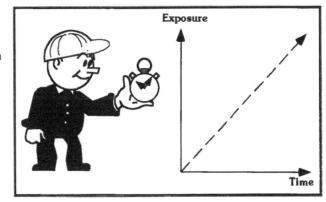

DISTANCE: Maximize the distance from a source of penetrating radiation

SHIELDING: Maximize the shielding between personnel and a source of penetrating radiation

FIGURE 8.1 TIME, DISTANCE, AND SHIELDING

8.3 PROTECTION FROM CONTAMINATION

The primary reason for controlling surface contamination is to minimize the possibility of inhalation or ingestion of radioactive material. In addition, surface contamination is controlled to reduce the possibility of any spread of radioactive material.

Loose surface contamination can easily be transferred to clothing, shoes, and hands or other surfaces of the body. Once contamination has been transferred to the hands, it is relatively easy for it to find its way to the mouth and be swallowed. Mechanisms by which contamination might find its way from the hands to the mouth include rubbing the face, handling food when eating, and smoking cigarettes.

There are several ways to control contamination. One of the most obvious and effective ways is to identify contaminated areas and make workers aware of the problem. Posting warning signs, barrier tape, or rope in contaminated areas will warn workers that the possibility of contamination exists. An area where loose surface contamination exists or is suspected to exist should be posted as a Restricted Area.

There are four common-sense rules for radiation protection when working in contaminated areas:

1. No eating, drinking, smoking, or chewing in a contaminated or restricted area. These are clear methods by which contamination can enter the body. The best precaution is not to take food, beverages, or cigarettes into a restricted area. This is usually mandatory.

2. Use respiratory protection equipment and protective clothing properly. Even the best respirator will not provide protection if worn incorrectly. Like the respirator, protective clothing must be worn in the intended manner if it is to provide adequate protection. Careful removal of protective clothing and personal decontamination are also important.

3. Always perform a frisk when exiting a contaminated area. The choice of hand and foot frisk or whole-body frisk is dependent upon the nature of the work being performed. This is largely common sense but may be specified by the radiation safety supervisor or the requirements of the radiation license. Frisking ensures that contamination is not removed from a restricted area and that personnel are not contaminated.

4. Employ good personal hygiene. Wash the hands and face before eating, drinking, or smoking. Keep nails trimmed and clean. Launder work clothing regularly. The reasons for all of these practices are so obvious that they need no explanation.

The frisk is performed using a Geiger-Mueller pancake probe or "frisker." The probe is held within 0.5 inches (1 centimeter) of the body and moved at a rate of 2 to 3 inches (5 to 8 centimeters) per second. If contamination is detected, the radiation safety officer should be contacted immediately to determine decontamination methods and to complete the necessary documentation.

Workers should think very carefully before scratching an itch or wiping sweat from an exposed face. Common sense is the key! In certain circumstances, it may be advisable to have a worker at the periphery of a restricted area who can perform such functions for workers who are badly contaminated.

Methods for personnel decontamination are described in Chapter 7, should they become necessary.

8.4 Protective Clothing

A layer of clothing can protect the wearer from beta radiation exposure to the skin. However, this is rarely a problem and is not an issue with NORM. Realistically, normal protective clothing cannot provide any direct protection from radiation exposure. Protective clothing is worn in contaminated or restricted areas to protect workers from contamination. It provides personnel with an easily removed outer "skin" so that if contamination is present on the clothing, it cannot cause continued exposure to the wearer once the clothing is removed. The other significant purpose of protective clothing is to eliminate a means of spreading contamination. When protective clothing is removed, the wearers are effectively decontaminating themselves.

At the daily pre-work briefing, a document (such as a Work Plan Briefing or Radiation Work Permit) detailing the planned work and associated precautions should be made available to workers. It should be read by each worker and/or read out loud by the work supervisor so that each worker is aware of the required protective clothing. It should also be checked for any "special" protective clothing that may be required. Workers will usually be required to sign the briefing document or a separate checklist, confirming that they are familiar with the required protective clothing and procedures. The radiological supervisor, at his or her discretion or as conditions dictate, may change protective clothing requirements. He or she should be contacted by workers if they have any questions or if problems arise.

8.4.1 Clothing Selection

The selection of protective clothing and the manner in which it is worn depend upon many factors, including the work being performed, the specific task of each worker, the weather, and the requirements of any licenses. Normal work clothes are suitable for some NORM work. However, there are several special considerations.

COVERALLS—When NORM workers are likely to become contaminated, the use of disposable coveralls should be considered. It is always best to select coveralls with no pockets. This avoids a snagging hazard and removes a potential mechanism for self-contamination. Pockets can (should) be sealed with tape if necessary. The use of disposable clothing can generate a substantial amount of potentially contaminated waste. Regular cotton coveralls may be used provided the NORM is not too wet. Daily laundering is then required, which necessitates a dedicated laundry system. In such a laundry setup, wastewater is recycled or stored and sampled before appropriate disposal.

BOOTS—Rubber boots are the preferred form of footwear because they are easy to decontaminate. The fine grooves found on the soles of some expensive boots tend to trap small particles and can be difficult to clean. For this reason, cheap boots with their heavy lug soles seem to be the best option.

GLOVES—For general skin protection, latex medical gloves are very effective. They are cheap and offer almost the same tactility as bare hands. While surprisingly robust, they cannot stand up to heavy work. When such work is performed, regular cotton work gloves can be placed over the latex gloves. However, they may become badly contaminated and need to be discarded or laundered (see previous comments regarding laundering). For greater durability and reusability, chemical-handling gloves are available in a wide range of styles and thicknesses. Dish-washing gloves may even be considered.

HEAD PROTECTION—It is always advisable to provide some form of contamination protection when there is any possibility of contaminating the hair. Where there could be splattering or spray, hooded coveralls may be favored. In many cases, a normal hard hat may be sufficient. In cases where a hard hat would normally be required, the need for or use of contamination protection should never supersede the use of other protective headgear. A hood can be worn over or under a normal hard hat. Long hair should be firmly secured under the protective headgear.

EYE AND FACE PROTECTION—Choices range from safety glasses to goggles, full face shields, or respirator face masks. Once again, selection of eye protection for radioactive contamination should not override other normal safety concerns. There is rarely any situation where it is appropriate to have no eye protection at all. As a minimum, safety glasses with side shields should be worn. In addition to mechanical protection from airborne radioactive material or flying debris, eye wear will provide protection from beta radiation, which is known to be linked to cataracts.

NATURE OF WORK—The requirements for a worker feeding NORM into a slurrying system will be very different from those for a surveyor performing a site inspection at an inactive facility. The latter may require nothing more than rubber boots and a few pairs of latex gloves in a plastic bag to wear while taking samples. The worker on the slurrying job may wear a slicker suit, heavy-duty rubber gloves, rubber boots, a face shield, and a respirator with duct tape sealing the wrists and ankles.

WEATHER—In warm weather, there is almost no type of protective clothing that will be comfortable for long periods while providing suitable protection to a worker. The work supervisor will have to make the best compromise appropriate to the circumstances. Break schedules may need to be tailored to meet the requirements of worker comfort and safety.

8.4.2 USING PROTECTIVE CLOTHING

Protective clothing will only protect the wearer in a fashion commensurate with the way in which it is put on, worn, and removed.

FIT—Protective clothing must be selected to ensure that it will fit properly. Clothing that is too large makes work difficult and tends to spread contamination because it may drag along the floor. Clothing that is too small does not provide adequate protection because as the worker moves, the joins where the clothing meets (i.e., points where coverall sleeves meet gloves) are pulled open, exposing the skin to contamination. If the clothing is too small, it could tear during heavy work activity. This is a common problem with disposable overalls, which frequently do not fit well and are somewhat flimsy. Protective clothing should be comfortable and provide sufficient mobility so that the worker is not hampered.

CONDITION—Prior to donning, each article of clothing should be inspected for holes, torn seams, broken zippers, etc. If defects are found, the article should be discarded and another item selected. Cut-off sleeves or open buttons and zippers are not acceptable because they compromise radiological protection.

SEALING GAPS—A hermetically sealed worker is totally protected from any loose contamination. However, the elaborate precautions taken by persons working with highly toxic chemicals or nuclear waste are not appropriate for NORM work. Despite the preceding statement, it is surprisingly easy to obtain a close approximation to a totally sealed suit. When appropriate for the nature of the work being performed, gaps and joins in protective clothing should be taped closed. When sealing sleeves and ankles, one-and-a-half turns of tape is usually sufficient. A tab at the end of the tape should be folded over to make the tape (and consequently the clothing) easier to remove in an emergency.

DAMAGE—Tears that occur during the course of a job should be patched or the garment removed and replaced, as appropriate. Duct tape is usually suitable for making temporary repairs to protective clothing. Precautions should be taken to prevent contamination of underclothing or skin while making repairs. If changing clothing, a full frisk should be performed when the old clothing has been removed.

8.4.3 REMOVING PROTECTIVE CLOTHING

Protective clothing must be removed carefully to prevent contamination of clean areas, the skin, and personal clothing worn underneath. To accomplish this, the following guidelines should be observed:

- Probably the best order in which to begin removing protective clothing is to remove headgear and respirator (if worn), followed by peeling off any tape seals. Gloves and coveralls can be removed in either order (using common sense).

- Caution should be exercised during clothing removal to prevent contaminating clean areas with loose contamination on the protective clothing that could be shaken or rubbed off during the removal process. Movements should be controlled in such a way that any spray of contamination is directed toward the contaminated area.

- To remove coveralls, unzip them and then carefully remove the arms from the sleeves and roll the coveralls down the body inside-out, being careful not to contaminate underlying skin or clothing.

- When removing shoe covers, frisking footwear, or scrubbing the soles of boots, it is easy to lose one's balance. This is hazardous in itself but also offers the opportunity to "dab" a clean foot back down in a contaminated area. A good technique to aid in maintaining balance is to bring one leg across in front, resting the back of the ankle or calf against the knee of the other leg. This will hold the sole of the foot in a convenient position. It is best to hold onto or lean against a rail, wall, or work colleague, as necessary.

- A rule of thumb to remember when removing protective clothing is to touch contaminated with contaminated and touch clean with clean.

8.5 PROTECTION FROM AIRBORNE RADIOACTIVE MATERIAL

There are many possible ways to generate airborne radioactive material. It can be the result of cutting or grinding operations on NORM-contaminated equipment. More commonly, airborne activity becomes a concern when handling dry materials such as pipe scale from tubulars or contaminated soil.

There are many means of dealing with airborne radioactive material, only one of which involves the use of respirators. Indeed, the use of respirators should normally be considered a last resort after other steps have been taken to reduce exposure to the airborne radioactivity. Such steps would generally be considered engineering controls.

The very first consideration in protection from airborne radioactive material is employing engineering controls to reduce or eliminate the amount of material becoming airborne in the first place. The most common means by which NORM becomes airborne are blowing dust and spray from high-pressure washing.

Dust may be almost completely eliminated by judiciously wetting the material in question. Wetting can usually be accomplished by careful use of an occasional gentle water spray. It is rarely advisable to completely soak the material because that will add to the volume of contaminated material and may actually generate a whole new waste stream.

When washing out vessels, the use of high-pressure water poses a tricky problem. When high pressures are used, a cleaning task can often be accomplished with a much smaller overall volume of water than would be used in low-pressure washing. This may offset the hazards of generating airborne radioactivity in the form of spray. The decision depends upon factors such as the activity of the material involved, the estimated total volumes of wastes and their nature, and the ability to recycle wash water. The decision may also be influenced by whether workers will be entering a vessel and must wear breathing apparatus anyway. If the decision is made to use high-pressure washing and protect personnel accordingly, using high-pressure steam, which will generate an even smaller volume of waste, should be considered.

Engineering considerations should also extend to a re-examination of the entire process that is generating the airborne radioactivity. For example, the cleaning of production tubing by traditional means could be replaced by a high-pressure water-blast process or could include a vacuum system. Many companies now have suitable equipment to perform cleaning in such a fashion. The use of enclosures to catch spray and systems to recycle waste-water is also a part of engineering control.

If engineering controls are not totally effective or are otherwise impractical, personnel exposures should be evaluated and considered. Exposures may be small enough that workers could not be expected to acquire a significant committed dose equivalent in the time they would be expected to be working. The following factors should be noted:

- Applying ALARA policy means that maintaining exposures to airborne radioactive contamination within legal limits is not sufficient if there are practical alternatives.

- Rotating workers through those tasks which result in exposure to the greatest concentrations of airborne radioactive contaminants will result in many workers receiving a small dose rather than one worker receiving a larger dose.

- It is commonly believed that the wearing of breathing apparatus is itself a health risk to workers. It causes an increased strain on the cardiopulmonary system and may also lead to overheating. For some tasks, it is also possible that wearing a respirator could limit the worker's vision, with corresponding safety implications. For this reason, respirators should never be used when they are not specifically required.

- When working in confined spaces, respiratory protection is often mandatory. In such cases, it should be used regardless of radiological conditions. However, it may be necessary to substitute filter cartridges that are approved for radionuclides.

- In most cases, it is the work itself which is responsible for creating the airborne radioactivity. This means that it is unlikely that the work supervisor will be able to assess the level of airborne activity until work has commenced. It then becomes a judgment call on the part of the supervisor as to whether or not work should begin with or without respiratory protection equipment being employed. Factors such as prior experience and radiological analysis of a sample of the material being worked with must be accounted for in making such a decision.

Legal limits on exposure to airborne radioactivity have two components: annual limits on intake (ALI) and derived air concentration (DAC). The ALI is the amount of a radionuclide which if taken into the body by inhalation or ingestion in 1 year would cause a committed dose equivalent of 5 rem (50 millisieverts) to the whole body or 50 rem (0.5 sieverts) to any one organ or tissue. In most cases, NORM workers are unlikely to receive a substantial dose due to airborne contamination for a significant number of days each year. Assuming that this is the case, the limiting factor to their allowed exposure is the DAC. A DAC is the concentration of a radionuclide which if inhaled by an average person under conditions of light work for 2000 hours per year would result in the intake of 1 ALI.

The work supervisor will normally be responsible for determining the best method of assessing airborne radioactivity and implementing appropriate protective measures. If the use of breathing apparatus is determined to be appropriate, workers cannot simply be handed respirators. A complete discussion of respirators and the rules and regulations governing their use is beyond the scope of this text. The basic requirements include:

- A respiratory protection program should be in place. This is a set of written procedures and practices which stipulate how, when, and what type of respirators will be used by the company and what other related measures will be taken. Such a program does not supersede the necessity to comply with the law.

- Personnel who will be wearing respiratory protective equipment need to be trained in the use of the chosen devices.

- All personnel who may be using respirators should also have passed a cardiopulmonary function test as part of a medical examination specifically intended to determine their fitness to use respiratory protection equipment.

- There are other specific requirements for certain types of breathing apparatus. Many types of breathing equipment must be fitted for the persons who will be using them. Each worker will probably prefer to have his or her own respirator, but if this is not the case, each must use the size and type of device for which he or she has passed a fit test. Half-face respirators should be tested every time they are donned. This can be accomplished using any of several commercially available "smelly" or irritant chemicals chosen according to the protective qualities the respirator is supposed to possess.

- The wearing of facial hair is incompatible with the use of most types of respiratory protection equipment. All personnel who may be required to wear such equipment must be clean-shaven.

- Different devices have different "protection factors" assigned to them. The protection factor of a given type of breathing device is an assigned factor by which, it is assumed, the respirator is capable of reducing airborne radioactivity. (For example, a half-face respirator, generally assigned a protection factor of 10, will allow a person to assume that he or she is only actually breathing in one-tenth or less of the radioactive material in the air.)

- Filter cartridges should be compatible with protection from radionuclides. This is usually stated on the cartridge itself. Such cartridges are usually color coded in purple or mauve.

- If using an external supply of air, care should be taken to ensure that the air supply is not tainted and any air intake is in an area with suitably low levels of airborne activity or is appropriately filtered.

- Note that no respiratory protection equipment can provide protection if it is not properly worn. Legally, it cannot be considered to provide any protection if the worker has not been fit-tested for that particular device.

8.6 Emergency Response

Emergencies involving radioactive materials generally fall into two categories:

1. Incidents in which the primary concern is that they could result in the release of radioactive material.

2. Incidents where human life, health, or property is threatened and radiological concerns are a secondary consideration.

In some industries where radioactive materials are encountered, accidents involving radiation and radioactive materials may constitute life-threatening emergencies because of the radiological hazards. Because this is unlikely to ever be an issue with NORM, the following general principle applies: **serious personnel safety concerns should always take priority over radiological concerns**.

8.6.1 Spill Response

In the event of a spill or accidental discharge of NORM waste, the work supervisor should ensure that all workers are familiar with the actions that should be taken. Such actions will often be incident-specific and should be discussed at work safety briefings. A general spill contingency plan will follow the basic procedures outlined below. (The acronym SWIM may be useful in remembering what steps should be taken.)

STOP THE SPILL—The primary effort of personnel upon noticing a spill should be to stop the spill or discharge by whatever means possible without endangering themselves or others.

WARN OTHERS—The first person who notices a spill should immediately warn other workers in the vicinity so that they may assist or avoid becoming contaminated, as appropriate. The work supervisor should also be notified as soon as possible. He or she will then take over responsibility for making additional notifications as required and will direct appropriate protective responses and cleanup actions.

ISOLATE THE AREA—Persons combating a spill should prevent non-involved persons or others who may be untrained in radiation safety precautions from entering the area of the spill. If a serious incident takes place, one person may be designated solely to perform this function. He or she should try to remain outside the contaminated area, if possible.

MINIMIZE EXPOSURE—During cleanup operations, ALARA principles must not be abandoned. Every effort should be made by all personnel to minimize their exposure to NORM contamination by wearing appropriate protective clothing and using reasonable care. Additional consideration should be given to the possibility of spilled or accidentally discharged waste further migrating due to the effects of rain or wind. Appropriate precautionary measures should then be taken.

Spill cleanups should take place as if they were a NORM-related job in themselves. All normal procedures and release requirements should be employed.

8.6.2 FIRE RESPONSE

In the event of a fire in which NORM material is burning or may become incorporated into a fire or in cases where the fire may cause the release or spread of NORM, the following factors should be considered:

- Fire-fighting and personnel safety will normally take precedence over any radiological controls.

- Upon discovery of a fire, all NORM-related operations should cease and unnecessary personnel should leave the area of the fire immediately.

- The work supervisor should be informed as soon as possible. It will then be his or her responsibility to coordinate activities and to call the fire service, if necessary.

- Emergency personnel attending a fire must be informed of the radiological hazards present and given whatever advice and assistance they require to safely conduct their jobs.

- Consideration should be given to the possibility that the use of large volumes of water to fight a fire may spread contamination. Other methods such as dry powder or CO_2 should be used if available and appropriate.

- The heat from a major fire could cause sealed drums containing NORM waste to rupture. Those containers most at risk should be either vented by puncturing them or sprayed with water to keep them cool.

- Should NORM-contaminated material be burning or otherwise incorporated into a fire, it should be assumed that there is a possibility of generating airborne radioactivity. Air monitoring should be initiated downwind of the fire, if it is possible to do so without endangering personnel or detracting from the fire-fighting efforts.

- If a fire occurs, once it is out or under control the work supervisor should assess the possibility that the fire may have caused any radioactive release. If there is a real possibility that this may have happened, then appropriate monitoring such as wipes or soil and water sampling should be conducted to assess the extent of contamination. Areas found to be contaminated should be secured and posted. Cleanup will be treated the same way as for a spill.

- Any personnel involved in fighting a fire or who could have been contaminated should be monitored and decontaminated if necessary, prior to leaving the site.

- No personnel should ever be required to jeopardize their own personal safety in order to prevent the spread of NORM contamination.

Chemical spills or other similar hazards that involve NORM-contaminated equipment or work areas should usually be treated in a similar fashion to a fire, given due regard for the nature of the hazards involved.

8.6.3 RESPONSE TO STORMS OR FLOODING

In the event of storms, hurricanes, flooding, or other natural hazards, personnel safety should take priority over radiological concerns. However, if it is possible and safe to do so, the following should be considered:

- NORM-contaminated equipment should be secured and sealed in an appropriate fashion to try to prevent loss or damage.

- Open drums or vessels containing NORM waste should be emptied into a secure container and sealed or covered to prevent NORM from being released due to high wind, flooding, or overflow with rain water.

- If serious flooding is likely to overwhelm a work site, then any NORM-contaminated equipment should be moved to a safe place. Equipment or containers that cannot be moved should be secured to reduce the possibility of being washed away.

8.6.4 Response to an Injury Accident

- Should an injury which requires immediate medical attention occur, that attention should be given without regard for the radiological consequences. Any resulting contamination can be dealt with in an appropriate fashion. Efforts should be made to minimize the possibility of spread of contamination only insofar as the treatment of the injured person is not compromised. Care should be taken to minimize the possibility of NORM-contaminated materials entering an open wound.

- A list of emergency contacts, including phone numbers of emergency services, radiological supervisors, and regulatory agencies, as well as directions to the nearest hospital, should be made available to all personnel working at a temporary job site. The whereabouts of this information and the nearest telephone or other communications device should be stated at each safety meeting.

8.7 Radiation Monitoring

Monitoring personnel for radiation dose does not seem to be a means of preventing radiation exposure. However, knowledge of the radiation dose that a person has received is an important part of any radiation protection program for the following reasons:

- It allows assessment of the health risk to which a person has been exposed.

- It provides reassurance (if not actual proof) of the effectiveness of the radiation protection program.

- To help to identify any radiological problems (whether actual incidents or not) which will assist in applying appropriate corrective measures.

- To show that workers have not received radiation exposures that exceed allowable regulatory limits.

- To keep workers informed of their radiation exposures.

Dose monitoring is broken down into two categories: (1) external whole-body dose or deep dose equivalent and (2) internal committed dose equivalent. These two doses are summed to provide the total effective dose equivalent for an individual. When appropriate, doses to individual organs may be assessed.

External deep dose equivalent is normally assessed by means of dosimetry monitoring devices, such as the TLD badge. This is the most commonly employed means of external dose assessment among workers in the oil industry. These doses can also be estimated from measurements taken during routine radiation exposure rate surveys and knowledge of an individual's work schedules. This approach is obviously very time consuming and is not particularly accurate. Should a person's dosimetry badge be lost, the dose he or she has received can be assessed by this means. However, it is usually more convenient and appropriate to assign that person the same dose as a colleague who has worked similar amounts of time in the same locations. If several people fit this description, the most conservative approach would be to use the highest dose that any of them had received.

Internal committed dose equivalents are based upon knowledge of the uptake of radioactive material that a person has experienced. For uptakes in the legal range, this is difficult to measure directly. Assuming that an effective radiation protection program has been in effect and the required measurements have been made, assessment is best achieved by inference from airborne radioactivity measurements.

Direct measurement of body burden or the total amount of a radionuclide present in a person's tissues can be made by means of bioassays or body scanning. Such measurements can only be made in specialist laboratories employing specialized techniques and are therefore prohibitively expensive for routine monitoring.

Bioassay techniques involve measuring the amount of radioactive material in urine, feces, exhaled breath, blood, or (occasionally) tissue samples. When trying to measure uptakes of NORM, radium-226 or any of its radium-228 daughters may be measured. Urinalysis assays or radon exhalation are the most common means employed.

A body scan is accomplished by placing a person in an ultra-low-background counting room or chamber and measuring the radiation coming from the body as if he or she were a giant laboratory sample. Very sensitive, large, expensive detectors must be used and long count times are often employed. Accumulations of individual radionuclides in specific tissues can be measured by means of carefully positioned detectors and gamma spectroscopic analysis.

Environmental monitoring for radiation exposure, soil and water concentrations, and airborne radioactivity can be conducted to assess the existing radiological conditions at a site or to monitor for the spread of contamination or increases in ambient radiation exposure levels due to a known contaminated site or storage facility. Information derived from such monitoring can be used to assess the risk to outside personnel and the environment arising from the presence of NORM.

8.8 DOSE LIMITS

Some of the most significant federal radiation dose limits are listed in Table 8.1. Note that the allowed dose to individual organs is greater than the allowed whole-body dose. Members of the general public and minors cannot legally receive the same doses as trained, monitored radiation workers. The allowable dose to a fetus has recently been decreased. In an attempt to prevent discrimination against pregnant women, the limit only applies to the fetuses of women who voluntarily declare to their employers that they are pregnant.

8.9 EDUCATION

For personnel working with or around radiation and radioactive materials, education in all the principles and practices of radiation protection from NORM provides all the following benefits:

- It is the best way to ensure that a person's exposure to radiation and radioactive materials remains as low as reasonably achievable.

- Personnel who are trained in radiation protection principles and practices are not only able to minimize their own radiological-related

TABLE 8.1 SUMMARY OF FEDERAL RADIATION DOSE LIMITS

Type	Limit	
	Sv	rem
Annual Occupational Dose Limits for Adults		
Total effective dose equivalent	0.05	5
Sum of deep dose equivalent and committed dose equivalent for any organ except the eye	0.5	50
Eye dose equivalent	0.15	15
Shallow dose equivalent to skin or extremity (extremities include knees and legs below the knee, elbows and arms below the elbow)	0.5	50
Annual Occupational Dose Limits for Minors 10% of the doses listed above		
Occupational Dose to an Embryo/Fetus The occupational dose to an embryo/fetus of a declared pregnant woman (a woman who has voluntarily informed her employer, in writing, of her pregnancy) may not exceed 5 mSv (0.5 rem) over the entire pregnancy		
Members of the Public The total effective dose equivalent that members of the public may receive as a result of licensed activities may not exceed 1 mSv (0.1 rem) annually or 0.02 mSv (2 mrem) in any one hour		

Note: The above list of dose limits is neither complete nor exhaustive. Reference should be made to the latest state and federal regulations for full, complete, and up-to-date requirements. Under certain special circumstances, occupational exposures may exceed the limits listed above. These are known as planned special exposures.

risk, but will also be best able to minimize the risk to their co-workers, members of the public, and the environment.

- Personnel who understand the relative risks of the radiation doses that they may receive will be more familiar with the real health risks and issues involved. They are therefore less likely to try to avoid radiation-related work.

- Workers are less likely to spread unnecessary fear or cause uninformed gossip among co-workers or members of the public.

- The risk of litigation as a result of workers having an unreasonable fear or misperception of the hazards involved in NORM-related work is reduced.

- Suitable training of personnel will meet regulatory radiation protection training requirements and reduce the chance of inadvertent violations of other radiation protection regulations.

- It is widely believed that better educated workers can perform their jobs more efficiently and will have better morale.

9

REGULATION
OF NORM

STUDY OBJECTIVES

This chapter will enable the student to:

- List the most important state and federal agencies that administer radiation regulations applicable to NORM.

- Understand what the term "agreement state" means.

- Explain what is covered in 10 CFR 20 regulations and name the federal agency associated with these regulations.

- Understand the requirements of OSHA hazard communication regulations.

- Understand the scope and extent of Department of Transportation regulations.

- List some of the oil-producing states with NORM regulations in effect.

9.1 REGULATIONS APPLICABLE TO NORM

The following pages contain a brief discussion of various federal and state regulations, including effective, pending, and proposed NORM regulations. The information presented is an outline only and is believed to be correct at the time of publication. Readers should contact the actual state authorities for complete and up-to-date copies of the regulations.

There are a surprising number of different regulations that apply to NORM. The most important ones can be divided into two main groups:

- Federal regulations concerning general radiation protection and their state equivalents.

- State regulations that apply specifically to NORM.

There currently are no federal regulations specifically for the control of NORM. Many states take care of radiation protection within their own borders rather than allowing the federal government to do so. They are known as "agreement" states. The general radiation protection standards that these states apply usually reflect, verbatim, the federal regulations. However, the states are free to enact their own regulations and do so for certain aspects of radiation protection.

Although NORM is generally not regulated as hazardous waste, exceptionally large or highly radioactive shipments of NORM waste can exceed the exemption levels set in Department of Transportation (DOT) Hazardous Material Transportation regulations (49 CFR 171-78) and thus become subject to the requirements of those regulations. Strictly speaking, DOT regulations normally only apply to interstate transportation. In practice, most states require that DOT regulations be followed at all times when transporting radioactive materials on a public highway. Check with state authorities to determine their policy on this.

As there are no federal NORM regulations, many states have either enacted their own or are considering doing so. States that have NORM regulations in effect at the time of this writing (in chronological order of promulgation) are:

- Louisiana

- Mississippi

- Arkansas

- Texas

- Georgia

- South Carolina

- New Mexico

Other states working toward or actively considering NORM regulations are:

- California

- Connecticut

- Illinois

- Kentucky

- New Jersey

- Oklahoma

- Washington

Some of these states already have NORM guidelines in effect.

In some of the states listed, oil field NORM is not the major concern, and regulations are aimed at fertilizer mining, phosphoric acid production, and other industries.

9.2 10 CFR 20

Part 10 of the Code of Federal Regulations, Section 20 (10 CFR 20) is the set of regulations produced by the Nuclear Regulatory Commission (NRC) that establishes basic federal standards for radiation protection. The regulations cover dose limits for workers and the public, dictate the use of the ALARA concept, set out monitoring requirements, and define the methods that must be used to quantify radiation exposures.

10 CFR 20 also contains tables of data which can be used in the calculation of exposures and regulatory limits for doses due to certain uptakes, ambient concentrations, and discharges. These tables are typically reproduced more or less unaltered in most state radiation protection regulations. They are quite lengthy and often constitute half of the entire radiation protection regulations. The detailed tables are based upon scientific data which describe the likely uptake mechanisms, biological processes, and organ-specific radiation doses associated with different physical and chemical forms of each of several hundred radionuclides.

Updated in 1992, these tables and the other dose limits contained in 10 CFR 20 represent the most current federally accepted standards for radiation protection. They are based upon recommendations from several sources, principally the International Commission on Radiological Protection (ICRP), United Nations Scientific Committee on the Effects of Atomic Radiation (UNSCEAR), National Council on Radiation Protection and Measurements (NCRP), U.S. Environmental Protection Agency (EPA), and the National Research Council Biological Effects of Ionizing Radiation committee reports (BEIR I through V).

It is in 10 CFR 20 that the dose limits quoted in Chapter 8 of this book are contained.

9.3 OSHA

The Occupational Safety and Health Administration (OSHA) has included standards for occupational radiation dose limitations in its regulations. These standards are the same as or similar to those of the NRC, contained in 10 CFR 20.

Of more interest to most persons responsible for radiation protection are the requirements of OSHA's "Right to Know" and "Hazard Communication" regulations. These regulations mandate that all persons who *could* receive a radiation dose (or who could be exposed to other workplace hazards) as a result of their employment be instructed as to the nature of the risk, its possible health effects, and appropriate protective measures. The importance of complying with these requirements cannot be overstressed. In today's litigious society, neglecting to provide adequate hazard awareness instruction can be (and has been) successfully used to litigate in cases where there was actually no significant radiation exposure. In such cases, the *perceived* risk (from the point of view of the litigant) has been more important than the actual risk because of the lack of proper instruction.

9.4 EPA

U.S. Environmental Protection Agency regulations apply primarily to environmental releases. Considerable attention has recently been paid by this agency to radon in homes and business premises. However, the EPA has also issued radiation protection guidelines aimed primarily at other federal agencies. Certain provisions in these guidelines have been adopted as part of regulations issued by several federal agencies.

Most notable of the EPA radiation protection guidelines are the requirements that no exposure should occur unless an overall benefit is derived from the activity causing the exposure and that radiation doses must be maintained as low as reasonably achievable (ALARA). Other requirements are in accordance with the NRC regulations for occupational and non-occupational doses and doses to minors and pregnant women.

The EPA has proposed cleanup rules for federal facilities. If a federal facility with radioactive soil contamination covered by this rule has NORM commingled with source and by-product material and the NORM contributes to the risk, it must be included in the cleanup. This is a somewhat limited coverage of the NORM issue; however, it is widely believed that these regulations could be adopted as the basis for NORM cleanup standards by many states.

9.5 DOT

Department of Transportation regulations for the transportation of radioactive materials have recently undergone a major revision. These regulations (contained in 49 CFR 171-78) are effective as of April 1, 1996. Of interest is that these regulations *require* the use of S.I. units after April 1, 1997. Identifying all the requirements of these regulations that could possibly be applied to NORM is beyond the scope of this book.

It should be noted that the majority of NORM shipments are exempt from most DOT regulations. Some NORM contractors have been in the habit of labeling drums of NORM waste as LSA (Low Specific Activity). Notwithstanding the fact that there are now three categories of LSA and that NORM could occasionally fall into the LSA I or conceivably LSA II category, most NORM shipments labeled as LSA have not met the criteria for such posting. Over-posting of radioactive material shipments is viewed almost as dimly as under-posting because it can lead to unnecessary concerns. In addition, an LSA shipment has other requirements apart from just posting and these are rarely complied with for NORM shipments. When NORM does become subject to DOT regulations, it most often falls into the "Limited Quantity" category.

9.6 STATE REGULATIONS

9.6.1 TEXAS

Texas may have the most confusing NORM regulations. Jurisdictional issues abound, with the Department of Health, Railroad Commission, and Natural Resources Conservation Commission (NRCC) each controlling some aspect of NORM regulation. In practice, almost all day-to-day issues fall under the Department of Health, Bureau of Radiation Control. The Railroad Commission regulates disposal of oil field wastes, including NORM, and the NRCC has responsibility (although no regulations at the time of this writing) for disposal of non-oil-field wastes, such as phosphogypsum.

Texas regulations use 50 μR/hr absolute (including background) as the exemption level for NORM-contaminated equipment and 5 pCi/g (0.19 Bq/g) of Ra-226 or Ra-228 as the exemption limit for soil and loose NORM material. The limit for layers of soil more than 15 cm below the surface is 15 pCi/g (0.56 Bq/g). If the radon emanation rate is less than 20 pCi/m²/sec (0.74 Bq/m²/sec), 30 pCi/g (1.1 Bq/g) may be used as the exemption limit. In practice, as in other states that have included this strange requirement (a holdover from the uranium mill-tailings controls upon which NORM regulations are based), most people have to work with the 5 and 15 pCi/g (0.19 and 0.56 Bq/g) limits, as it is often unrealistically difficult to determine a radon emanation rate. The exception is for underground disposal where radon emanation is irrelevant and the 30 pCi/g (1.1 Bq/g) limit may be used.

Texas allows on-site disposal by mixing with the land surface providing the resulting NORM concentration is less than 5 pCi/g (0.19 Bq/g) of Ra-226 or Ra-228 above background.

9.6.2 LOUISIANA

Most of the states with NORM regulations are, at best, flirting with their first round of revisions. Louisiana, on the other hand, is now on its third significant version. Louisiana was the first state to introduce NORM regulations and has led the way in enforcement and management practices.

In similar fashion to Texas, Louisiana has separate jurisdictions over general NORM management and handling and underground NORM disposal. Injection disposal of NORM is forbidden. However, disposal into wells during abandonment requires approval from both the Department of Environmental Quality, Radiation Protection Division and the Office of Conservation, Injection and Mining Division.

The exemption limit for contaminated equipment is 50 μR/hr absolute. Soils and loose NORM are regulated at levels of 5 pCi/g (0.19 Bq/g) of Ra-226 or Ra-228 or 15 pCi/g (0.56 Bq/g) below 15 cm depth. Louisiana has dropped the radon emanation requirement in favor of allowing 30 pCi/g (1.1 Bq/g) as a cleanup standard when it can be shown that the total effective dose equivalent to the public will not exceed 100 mrem (1 mSv).

The second version of the Louisiana implementation guide for the management of NORM should be available early in 1996.

9.6.3 MISSISSIPPI

Responsibility for NORM in Mississippi is divided between the Department of Health and the Oil and Gas Board. The Oil and Gas Board has authority for NORM at the well site, while the Department of Health has jurisdiction once NORM is removed from its place of generation. This is a recent reorganization. Prior to July 1995, the Department of Health had sole responsibility for NORM.

The proposed regulations of the Oil and Gas Board are less stringent in certain areas than the Department of Health regulations that previously covered well site NORM and as a result are being strongly contested by certain members of the legal profession who have been leading their field nationwide in NORM-related litigation. Although NORM-related class-action suits and similar litigation have been seen in several states, Mississippi has a disproportionately high incidence of such activities.

Exemption limits for NORM in Mississippi are 25 μR/hr above background for contaminated equipment and 5 pCi/g (0.19 Bq/g) over the first 15 cm of depth for soil, with 15 pCi/g (0.56 Bq/g) thereafter unless the radon emanation rate is below 20 pCi/m^2/sec (0.74 Bq/m^2/sec), when 30 pCi/g (1.1 Bq/g) may be used.

Some of these requirements may change if the new Oil and Gas Board regulations are adopted. As with all other regulatory information contained herein, the reader is referred to the relevant state authorities for up-to-date information on Mississippi NORM regulations and policy.

9.6.4 ARKANSAS

At the time of this writing, the exemption limit for NORM-contaminated equipment in Arkansas is 25 μR/hr above background. There is a proposal to change this to 50 μR/hr absolute. Arkansas limits for exemption and release of soils make no allowance for radon emanation and therefore stand at 5 pCi/g (0.19 Bq/g) over the first 15 cm of depth and 15 pCi/g (0.56 Bq/g) thereafter.

9.6.5 GEORGIA

Georgia has no oil and gas production. The state's NORM regulations, while broadly similar to those of oil-producing states, are intended to cover errant NORM materials brought in from other states and phosphogypsum mining and processing. The exemption limits are 50 µR/hr absolute for contaminated equipment and 5/15 pCi/g (0.19/0.56 Bq/g) of Ra-226 or Ra-228 in soils, with a possible 30 pCi/g (1.1 Bq/g) limit if radon emanation rates do not exceed 20 pCi/m^2/sec (0.74 Bq/m^2/sec).

9.6.6 NEW MEXICO

As it is in many other states, authority for regulation of NORM in New Mexico is divided between the Oil Conservation Commission, which has jurisdiction over underground disposal into injection wells or abandoned wells, and the Environment Department, Hazardous and Radioactive Materials Division, whose regulations cover the remaining NORM issues. Radioactive Materials Division regulations make numerous references to the division's NORM worker protection and operations guidelines, which have not been finalized as of this writing. Also, the Oil Conservation Commission has not finalized its regulations yet. The reader is referred to the New Mexico authorities for the latest information.

New Mexico's regulations are slightly different than those of other states in that radium-228 is ignored. This actually makes a lot of sense, as Ra-228 levels are usually less than those of Ra-226, and even when this is not the case, the two never differ by a significant proportion. Other common-sense approaches include titling the regulations "Naturally Occurring Radioactive Materials (NORM) in the Oil and Gas Industry." While the dedication of NORM regulations to oil and gas production waste is implied in other state regulations, it is not categorically stipulated.

Radiation exposure rate exemption limits in New Mexico are 50 µR/hr absolute for contaminated equipment and removable sludges and scales contained within that equipment. A concentration exemption and release limit of 30 pCi/g (1.1 Bq/g) of Ra-226 (with no depth, radon emanation, or dose qualifiers) applies to contaminated soil and to removable sludges and scales.

9.6.7 SOUTH CAROLINA

Like Georgia, South Carolina has no oil and gas production. Similarly, NORM regulations are intended to cover errant NORM materials brought in from other states, phosphogypsum mining and processing, and other industrial wastes. The exemption limits are 50 μR/hr absolute for contaminated equipment and 5/15 pCi/g (0.19/0.56 Bq/g) of Ra-226 or Ra-228 in soils, with a possible 30 pCi/g (1.1 Bq/g) limit if radon emanation rates do not exceed 20 pCi/m^2/sec (0.74 Bq/m^2/sec).

South Carolina is the only state to use the term TENR (which is more technically correct and stands for Technologically Enhanced Natural Radioactivity) instead of NORM.

9.6.8 OTHER STATES WITH NORM CONTAMINATION AND/OR PENDING REGULATIONS

Other interesting and significant NORM regulatory information, alphabetically by state, includes:

ALASKA has NORM problems but no regulations. The state should start the process of drafting NORM regulations some time in 1996.

CALIFORNIA does not have a very severe NORM problem in the oil industry, compared to other major producing states (although geothermal power production facilities have an incredible scale buildup problem with attendant NORM issues). However, the state has been looking at possible NORM regulation for quite some time. There are no immediate plans to put regulations into effect.

COLORADO has a moderately sized oil industry with a modest NORM problem. NORM regulations have been proposed, but no timetable for implementation is known at the time of publication.

FLORIDA has a small oil industry and a large phosphogypsum problem. At the time of publication, the state has not decided whether NORM regulations are necessary.

HAWAII, surprisingly, has a small amount of NORM produced as a result of geothermal energy. There is no urgency to address NORM, but an upcoming sweeping revision of the state's radiation protection regulations may include it anyway.

ILLINOIS has draft NORM regulations undergoing final revisions but no timetable for introduction.

KANSAS has a moderately sized oil industry but is only slowly moving to address the possible need for NORM regulations.

KENTUCKY is a minor oil producer with no firm plans on NORM regulation. However, the state authorities have been working with some oil operators on the evaluation and remediation of contaminated properties.

MICHIGAN has had draft NORM regulations circulating for several years but apparently has no firm plans to finalize them or put them into effect.

NEW JERSEY is not an oil-producing state, but the Division of Environmental Quality believes that water treatment and certain other industrial operations in the state can produce NORM at sufficient levels to be of concern. This is due to the high radon content of most of the state's water supplies. New Jersey has proposed NORM regulations in circulation, which will be put into effect as soon as the revision process is completed.

NEW YORK has very little oil production but has included NORM in its regulations for radioactive waste disposal and soil cleanup.

NORTH DAKOTA, although an oil-producing state with some evidence of NORM production, has draft regulations but no timetable for adoption.

OKLAHOMA has a large oil industry and proposed NORM regulations undergoing final revisions pending adoption.

OREGON's oil industry is small, but it has some NORM associated with Zircon sand (a mineral extraction by-product). At the time of publication, the state is not working on regulations.

PENNSYLVANIA has no NORM regulations planned, despite having a moderate oil industry.

WASHINGTON STATE has recently enacted regulations limiting the disposal of NORM waste at Low Level Radioactive Waste disposal facilities. In practice, this only affects one disposal facility whose operator is attempting to overturn the rule.

WISCONSIN has some regulations governing disposal of materials containing radium-226, but they do not really address NORM per se.

9.7 OTHER NORM REGULATION ISSUES OF INTEREST

The Minerals Management Service (MMS), which regulates oil and gas production in federal waters, has regulations primarily governing NORM disposal.

The Conference of Radiation Control Program Directors (CRCPD), an organization made up of representatives from all U.S. radiation regulatory bodies, produces guidelines and suggested regulations on a wide range of radiological topics by means of a vast network of committees. The CRCPD is currently working on suggested NORM regulations.

The Canadian province of Alberta has a comprehensive set of voluntary NORM guidelines dealing with its considerable NORM problems related to both oil production and fertilizer manufacture, but no regulations.

The United Kingdom was the first country to develop and implement NORM regulations, during the 1970s. Several other countries with oil- and gas-producing operations in the North Sea and elsewhere in Europe have regulations or guidelines in place.

Almost all oil-producing countries have NORM problems due to oil and gas production. Many countries have major problems related to mining, chemical and fertilizer production, and other industries. For the majority of countries with the worst problems (the countries of the former Soviet Union for example), NORM controls are a very low priority. Some of the Arabian countries are believed to be considering implementing NORM control programs and regulations.

10 GLOSSARY OF TERMS

ACTIVATION—The process whereby inactive material is made to be radioactive. Usually by exposure to neutron radiation.

ACTIVITY (RADIOACTIVITY)—The number of disintegrations in a radioactive substance per unit time. Units of curies, becquerels, or disintegrations per second (dps).

ACUTE (RADIATION) EXPOSURE—Exposure to radiation lasting for a short time period (usually less than 24 hours).

AIRBORNE RADIOACTIVITY AREA—Special type of restricted area, access to which is controlled for the purposes of controlling exposure to airborne radioactive material.

ALARA—An acronym for "as low as reasonably achievable," a key concept in radiation protection. Based upon the premise that any exposure to radiation, no matter how small, has some health risk associated with it.

ALI (ALLOWABLE LIMIT ON INTAKE) REGULATORY LIMIT—The amount of a radionuclide which if taken into the body by inhalation or ingestion would cause a committed dose equivalent of 5 rem to the whole body or 50 rem to any one organ or tissue.

ALPHA PARTICLE (α)—Type of radiation given off by NORM. Particle consisting of two neutrons and two protons. Possesses an electrical charge of +2e and has an atomic mass number of 4. Helium nucleus.

ALPHA RADIATION—Ionizing radiation consisting of alpha particles emitted by certain radionuclides during radioactive decay. Alpha radiation has greater ionizing ability but less penetrating ability than beta or gamma radiation.

ATOM—The smallest indivisible part of an element that can take part in a chemical reaction.

ATOMIC—Of, relating to, or comprising atoms.

ATOMIC MASS—The mass of an atom or sub-atomic particle. Measured in grams or atomic mass units (amu).

ATOMIC MASS NUMBER (A)—The number of nucleons in the nucleus of an atom. It is the whole number nearest to the atomic mass of the atom, measured in atomic mass units. Sometimes simply called mass number.

ATOMIC MASS UNIT (amu)—A unit of mass, used to express atomic and molecular weight. Equal to one-twelfth the mass of an atom of carbon-12 or 1.66×10^{-4} grams. Also called unified atomic mass unit or dalton.

ATOMIC NUMBER (Z)—The number of protons in the nucleus of an atom. Defines it as being a certain element.

BACKGROUND RADIATION—Ambient environmental radiation levels due to causes other than the specific source being measured.

BECQUEREL (Bq)—S.I. unit of activity equal to 1 decay per second (dps).

BETA PARTICLE (β; SOMETIMES β^- OR β^+)—Electron (or occasionally positron) emitted by certain atoms during radioactive decay. Electrical charge of –e (electron) or +e (positron).

BETA RADIATION—Ionizing radiation consisting of beta particles. Lower penetrating ability than gamma radiation but greater than alpha radiation.

BOHR MODEL—Simple model for the atom which describes the arrangement of its constituent particles in a manner consistent with the explanation of most atomic properties.

BREMSSTRAHLUNG—Electromagnetic radiation produced by the rapid acceleration (or deceleration) of a charged particle. Usually encountered as the mechanism for production of "secondary" x-rays by beta radiation.

CALIBRATION—Normalization of the response of an instrument to ensure that its output readings accurately reflect the magnitude of the properties being measured.

CATHODE—Negative electrode.

CHRONIC (RADIATION) EXPOSURE—Continuous or repeated exposure to radiation over a long time period.

COMMITTED DOSE EQUIVALENT—The total radiation dose to a part of the body that will result from retention of radioactive material in the body. For the purposes of estimating the dose commitment, it is assumed that from the time of intake, the period of exposure to retained material will not exceed 50 years.

CONTAMINATION, RADIOACTIVE—Radioactive material in any undesired location, especially where it may be an environmental or health hazard.

COSMIC RADIATION—Radiation originating outside the earth. Either solar (from the sun) or galactic (from the far reaches of the universe). Consists of charged particles (mainly protons) and electromagnetic radiation (x-rays) with a wide range of energies.

COULOMB FORCE—Force that acts between two electrically charged objects. Repulsive for like charges and attractive for opposite charges.

CURIE—Unit of radioactivity. An amount of radioactive material corresponding to 3.7×10^{10} disintegrations per second. Roughly equivalent to the activity of 1 gram of radium.

DAC (DERIVED AIR CONCENTRATION)—Measure of airborne radioactive material. The concentration of a radionuclide which if inhaled by an average person under conditions of light work for 2000 hours per year would result in the intake of 1 ALI.

DAUGHTER PRODUCT—Any nuclide that originates from a given radionuclide, the parent, by radioactive decay.

DECAY, RADIOACTIVE—The transformation of a radionuclide into a different nuclide or energy state by the spontaneous emission of particles or energy.

DETECTOR, GEIGER-MUELLER—Gas-filled radiation detection device. Operates at a voltage just below that which would cause continuous discharge between its electrodes. The voltage is sufficient to cause a multiplicative effect that results in a momentary discharge of the detector for every ionizing incident. The device is therefore used to count ionization events. Hence the nomenclature Geiger counter.

DETECTOR, SCINTILLATION—A device that enables the measurement of radiation by measuring the intensity of light or counting flashes of light induced in certain materials by the deposition of energy from radiation. The light flashes are detected by a photomultiplier tube which converts them into an electrical signal and then amplifies that signal. The magnitude of the electrical signal is dependent upon the energy deposited by the radiation. Therefore, scintillation detectors may be used for energy spectroscopy, which enables identification of the radionuclide that emitted the radiation.

DNA (DEOXYRIBONUCLEIC ACID)—A component of living cells that encodes all hereditary characteristics. It consists of two chains of alternating phosphate and sugar (deoxyribose) units wound into a structure known as a double helix. The chains are joined together by special bonds between complimentary base pairs (adenine and thymine or cytosine and guanine). The sequence of sugar/phosphate/base (nucleotide) groups known as genes determines all hereditary characteristics of the organism.

DOSE—A quantity of radiation or absorbed energy. Units are the rad or gray (Gy).

DOSE EQUIVALENT—A measure of radiation used for protection purposes. Intended to normalize dose to biological tissue for different types of radiation with different energies. Obtained by multiplying dose by a quality factor for the radiation in question. Units are the rem (roentgen equivalent man) and sievert. Usually (incorrectly) simply referred to as dose.

DOSIMETER—An instrument used to measure accumulated radiation exposure. Most often a passive device. Usually designed so that dose equivalent due to gamma, x-ray, beta, and often neutron radiation may be determined.

ELECTROMAGNETIC RADIATION—Non-particulate form of radiation. Waves of energy traveling at the speed of light. Characterized by its wavelength or frequency. Energy range includes gamma and x-rays, ultraviolet and visible light, infrared, and radio waves.

ELECTRON (e)—Sub-atomic particle. Negatively charged with a rest mass of 9.11×10^{-31} kilograms.

ELECTRON VOLT—Unit of energy commonly used to describe the energy of radiation. Magnitude equal to the energy gained by an electron when accelerated through a potential of 1 volt.

ELEMENT—One of 112 known types of atoms characterized by their atomic number.

EXPOSURE—The amount of radiation to which an individual is exposed. Unit of measurement is the roentgen.

EXPOSURE RATE—Radiation exposure per unit time.

FALLOUT—Deposition of airborne, man-made radioactive materials resulting from the detonation of a thermonuclear device or accidental emission from nuclear power generation or fuel reprocessing facilities.

FREE RADICAL—Chemically reactive atom or molecule characterized by having at least one unpaired electron. Often produced directly or indirectly by ionizing radiation.

FRISK—Informal term for personal survey to determine the presence or extent of contamination on the skin or clothing.

GAMMA RAY (OR RADIATION)—Ionizing electromagnetic radiation. Originates as excess energy from nuclear reactions.

GAS AMPLIFICATION CURVE—Graph that describes the relationship between the magnitude of the electrical signal from a gas-filled detector as a function of the voltage applied across its electrodes.

GEIGER COUNTER—Common name given to a gas-filled detector operating in the Geiger-Mueller region. *See* Detector, Geiger-Mueller. Frequently incorrectly applied to all transportable radiation detection instruments.

GRAY—S.I. unit of absorbed dose. 1 gray = 100 rad.

HALF-LIFE, BIOLOGICAL—The time required for the body to eliminate half of any radioactive material ingested by natural biological means.

HALF-LIFE, EFFECTIVE—The time required for half of a specific radionuclide present in the body to be reduced by one-half as a result of the combined action of radioactive decay and biological elimination.

HALF-LIFE, RADIOACTIVE—The time required for the activity of a given amount of any radionuclide to decrease by half due to radioactive decay.

INTERNAL EXPOSURE—Irradiation by radioactive material inside the body.

ION—An atom or molecule that has a net electrical charge as a result of the number of electrons and the number of protons being different.

IONIZATION—The process of producing electrically charged ions by breaking up electrically neutral atoms or molecules or by removing or adding electrons to same.

IONIZING RADIATION—Radiation capable of imparting sufficient energy to matter to cause ionization.

ION PAIR—A closely associated positive ion and negative ion having charges of the same magnitude and formed from a neutral atom or molecule by ionizing radiation.

ISOTOPES—Atoms of the same element that have an equal number of protons but differing numbers of neutrons. (Same atomic number "Z" but differing atomic mass number "A")

MICROROENTGEN—A unit of radiation exposure in common usage, equal to one-millionth of a roentgen.

MILLIREM—A unit of radiation exposure in common usage, equal to one-thousandth of a rem.

MOLECULE—The smallest part of a substance which is made up of two or more atoms that can express the definitive physical and chemical properties of that substance.

NATURALLY OCCURRING RADIOACTIVE MATERIAL (NORM)—Radioactive material naturally present in the environment that has been unintentionally concentrated. In an oil field, radium and radon and their daughter nuclides, concentrated as a by-product of oil and gas production.

NEUTRON (n)—Sub-atomic constituent of an atomic nucleus. Possesses no electrical charge. Rest mass slightly greater than that of a proton at 1.675 $\times$ 10^{-27} kilograms.

NORM—Acronym for Naturally Occurring Radioactive Material.

NUCLEAR—Of, or pertaining to, an atomic nucleus.

NUCLEAR FORCE—Strong, charge-independent, short-range force responsible for holding the nucleus of an atom together.

NUCLEI—Plural of nucleus.

NUCLEUS—The most massive part of an atom. Contains protons and neutrons. In a Bohr atom, the nucleus is considered to be at the center of the atom, surrounded by orbiting electrons.

NUCLIDE—Any atom, characterized by its atomic number, atomic mass, and energy state. Radioactive atoms are referred to as radionuclides.

PERIODIC TABLE—Arrangement of atomic elements in order of increasing atomic number and grouped according to their chemical properties.

PHOTOMULTIPLIER—Device usually encountered as part of a scintillation detector which produces an electrical output in response to light. The output has much greater energy than the light being measured.

PROTON (p)—Sub-atomic constituent of an atomic nucleus. Possesses a positive electrical charge of magnitude e. Rest mass slightly less than that of a neutron at 1.673×10^{-27} kilograms.

QUALITY FACTOR—A proportionality factor that relates radiation damage to body tissue to the actual radiation energy absorbed. Dependent upon the type and energy of the radiation.

RAD—Unit of radiation dose. The amount of radiation that will deposit 100 ergs of energy per gram of material.

RADIATION—Particles or electromagnetic waves that carry energy.

RADIOACTIVITY—The spontaneous disintegration of the nucleus of an unstable atom that results in the emission of radiation. *See also* Activity.

RADIONUCLIDE—A radioactive atom characterized by its atomic number, atomic mass, and energy state.

REM (ROENTGEN EQUIVALENT MAN)—Unit of dose equivalent. The amount of any type of radiation that will cause damage to body tissue equivalent to that which would be caused by absorbing 100 ergs of gamma ray energy per gram of body tissue.

RESTRICTED AREA—Any area to which access is controlled for the purposes of controlling exposure to radiation or radioactive materials.

ROENTGEN (RÖNTGEN) (R)—Unit of radiation exposure. The amount of radiation that will produce a charge of 2.58×10^{-4} coulombs on all the ions of one sign in 1 kilogram of dry air under standard conditions. Used for gamma and x-rays with energies below 4 mega electron volts.

SCALE—Hard deposit found on the inside of production tubing and other equipment. Deposited as a result of dissolved minerals coming out of solution. Relevant to radiation protection because it often contains NORM radionuclides.

S.I. UNITS (SYSTÈME INTERNATIONALE D'UNITÉS)—The internationally agreed upon system of measurement. Used for all scientific and most technical measurements. Still only partially accepted for radiation protection in the United States.

SIEVERT—S.I. unit of dose equivalent. Equal to 1 joule per kilogram; 1 sievert = 100 rem.

TENR—Acronym for Technologically Enhanced Radioactive Material. *See* NORM.

TERRESTRIAL RADIATION—The portion of background radiation that originates from naturally occurring radionuclides present on earth.

THRESHOLD DOSE—The minimum dose of radiation believed to be required to cause an adverse health effect. It is not widely accepted that all biological effects of radiation have a threshold.

TLD (THERMOLUMINESCENT DOSIMETER)—Dosimetry device which allows measurement of dose by absorbing and "storing" energy deposited by radiation. The energy is released in the form of a measurable flash of light upon heating.

X-RAY—Form of electromagnetic radiation generated by the acceleration of charged particles. Essentially identical to gamma radiation but covers a much wider energy range.

INDEX

139